Random Number Generators

Luis Gerardo de la Fraga •
José David Rodríguez-Muñoz •
Esteban Tlelo-Cuautle

Random Number Generators

Verilog Description, Hardware Implementation and Applications

Springer

Luis Gerardo de la Fraga
Department of Computer Science
CINVESTAV-IPN
Mexico City, Mexico

José David Rodríguez-Muñoz
Department of Electronics
INAOE
Tonantzintla, Puebla, Mexico

Esteban Tlelo-Cuautle
Department of Electronics
INAOE
Tonantzintla, Puebla, Mexico

ISBN 978-3-031-82867-6 ISBN 978-3-031-82865-2 (eBook)
https://doi.org/10.1007/978-3-031-82865-2

This Springer imprint is published by the registered company Springer Nature Switzerland AG
The registered company address is: Gewerbestrasse 11, 6330 Cham, Switzerland

Luis Gerardo de la Fraga wants to dedicate this book to his family. To all the time and love that we share and permits the time to write books.

José David Rodríguez-Muñoz wants to dedicate this book to his family and friends, who encouraged and supported him unconditionally throughout this project. Their encouragement, friendship, and company were fundamental to the completion of this work.

Esteban Tlelo-Cuautle wants to express his gratitude to his beloved family.

Preface

Random number generation (RNG) is a key technology that is used for information security in various fields such as electronic commerce and authentication. In addition, random numbers are used in various applications such as in the generation of keys for data encryption, games, lotteries, sampling, simulations, statistical sampling, search/sort algorithms, and gambling. The classification of RNGs encompasses linear and nonlinear (chaotic) pseudo and truly random number generators, and they can be evaluated by applying statistical tests, and compared based on speed and silicon area.

A pseudo-RNG (PRNG) requires a seed, which is a series of bits, which also determines its stability. Random numbers generated at a sufficiently long length can encrypt sensitive data and make it difficult for another computer or person to decrypt the data. In this manner, the challenge is the hardware implementation of a method of generating random numbers reliably. This book exploits the advantage of Field Programmable Gate Arrays (FPGAs) as reconfigurable hardware systems, to perform fast prototyping of chaos-based PRNGs. The FPGA-based PRNGs are based herein on chaotic maps and integer/fractional (hyper)-chaotic systems. Implementation details are provided from Verilog descriptions, FPGA synthesis and applications of authentication of encrypted RGB images and text. The hardware implementations are efficient and can be used for security, authentication, and Internet of Things applications.

Mexico City, México — Luis Gerardo de la Fraga
Tonantzintla, Mexico — José David Rodríguez-Muñoz
Tonantzintla, Mexico — Esteban Tlelo-Cuautle
2025

Preface

Acknowledgments

José David Rodríguez-Muñoz thanks CONAHCyT-Mexico for the scholarships to pursue a master of science degree at Instituto Nacional de Astrofísica, Optica y Electrónica (INAOE).

Esteban Tlelo-Cuautle thanks CONAHCyT/Mexico for the support for a sabbatical leave at CINVESTAV-Mexico through the program: Apoyos complementarios para estancias sabáticas vinculadas a la consolidación de grupos de investigación 2023.

Contents

1 **Introduction to the Design and Implementation of PRNGs Based on Chaotic Maps and Systems** ... 1
1.1 Introduction ... 1
1.2 True and Pseudo-Random Number Generators ... 3
1.3 Chaos and Cryptography ... 5
1.4 Generation of Pseudo-Random Numbers with Available Software ... 6
1.5 Hardware Implementation of Pseudo-Random Number Generators ... 9
1.6 Conclusion ... 12
References ... 12

2 **Numerical Methods to Approximate Integer-/Fractional-Order Chaotic Systems** ... 17
2.1 Introduction ... 17
2.2 Numerical Methods for Integer-/Fractional-Order Chaotic Systems ... 18
2.2.1 Numerical Methods for Integer-Order Chaotic Systems ... 19
2.2.2 Numerical Methods for Fractional-Order Chaotic Systems ... 20
2.3 Chaotic Systems with Self-Excited and Hidden Attractors ... 23
2.3.1 Chaotic Systems with Self-Excited Attractors ... 23
2.3.2 Chaotic Systems with Multistability and Hidden Attractors ... 25
2.4 Fractional-Order Chaotic Systems ... 28
2.5 Conclusion ... 32
References ... 32

3 **Verilog Descriptions of Digital Blocks to Synthesize Chaotic Maps and Systems** ... 35
3.1 Introduction ... 35

3.2 Verilog Descriptions of Arithmetic Blocks 37
3.2.1 Parallel-Input, Parallel-Output (PIPO) Register 38
3.2.2 Adder and Subtractor 40
3.2.3 Multiplication Block 43
3.2.4 Description of a Single Constant Multiplier 45
3.3 Verilog Descriptions of Other Combinational and Sequential Blocks 47
3.3.1 Multiplexer 47
3.3.2 Decoder 48
3.3.3 Type-D Flip-Flop with Enable Pin 49
3.3.4 Ascending Binary Counter 50
3.3.5 Bidirectional Binary Counter 52
3.3.6 Finite-State Machine 54
3.3.7 Random Access Memory (RAM) 56
3.3.8 Read-Only Memory (ROM) 57
3.3.9 Register-Based Memory 58
3.3.10 Multiplier Accumulator 60
3.3.11 Multiplier-Multiplexed 62
3.4 Verilog Description of a 2D Chaotic Map and 3D Lorenz System 64
3.4.1 Verilog Description of the 3D Lorenz System 65
3.4.2 Fixed-Point Notation for Lorenz System 65
3.4.3 Block and Verilog Descriptions of the Constants Multiplying Variables for Lorenz System 66
3.4.4 Block and Verilog Description of the 3D Lorenz System 71
3.4.5 Block and Verilog Descriptions of a 2D Chaotic Map 73
3.4.6 Selection of the Fixed-Point Format 73
3.4.7 Block and Verilog Descriptions of the Control of the Equations of the 2D Sprott Map 74
3.4.8 Block and Verilog Description of the 2D Sprott Map 78
3.5 Conclusion 80
References 80

4 Statistical Tests for PRNGs 83
4.1 NIST Randomness Test Suite 83
4.1.1 How to Use NIST Test Suite 83
4.2 TestU01 Test Suite 87
4.2.1 TestU01 Test Suite on Floating Point Numbers in [0, 1) 88
4.3 Dieharder Test Suite 91
4.4 Conclusion 92
References 92

5 PRNGs Based on Chaotic Maps and 3D, 4D, and 5D Chaotic Systems 95
5.1 Introduction 95

5.2 PRNG Based on a 2D Sprott Map 97
5.3 FPGA Synthesis of a PRNG Based on 2D Sprott Map 100
5.3.1 PRNG from Chaotic Time Series by Applying mod256 101
5.3.2 Verilog Description of a PRNG Based on the 2D Sprott Map 102
5.3.3 PRNG Based on 2D Chaotic Map: FPGA Implementation and Experimental Results 105
5.4 PRNG Based on Chaotic Maps Without Fixed Points 107
5.4.1 PRNG Based on the Chaotic Map Without Fixed Points nfp2 108
5.4.2 PRNG Based on the Chaotic Map Without Fixed Points nfp1 111
5.4.3 PRNG Based on the Chaotic Map Without Fixed Points nfp1 Implemented with Floating-Point Numbers 114
5.5 PRNG Based on the 3D Lorenz System 117
5.5.1 Verilog Description of a PRNG Based on the 3D Lorenz System 119
5.5.2 FPGA Implementation and Experimental Results 123
5.6 PRNG Based on a 4D Hyperjerk System 125
5.6.1 Time Series, Attractors, and Fixed-Point Format 126
5.6.2 Block and Verilog Description of a PRNG Based on the 4D Hyperjerk System 127
5.6.3 PRNG Based on a 4D Hyperjerk System: FPGA Implementation and Experimental Results 132
5.7 PRNG Based on a 5D Chaotic System 133
5.7.1 Time Series, Attractors, and Fixed-Point Selection 135
5.7.2 Block and Verilog Description of a 5D Chaotic System 135
5.7.3 PRNG Based on a 5D Chaotic System: FPGA Implementation and Experimental Results 139
5.8 Conclusion 143
References 143

6 PRNG Based on Fractional-Order Chaotic Systems 145
6.1 Introduction 145
6.2 Block Description of a Fractional-Order 1D Memristive Time-Delay Chaotic System 147
6.2.1 Selection of the Fixed-Point Format 148
6.2.2 Block Description of the Fractional Memristive System 148
6.3 Verilog Descriptions of the Fractional-Order Memristive System with Time Delay 150
6.3.1 Verilog Description of the Memory_XT Block 150
6.3.2 Verilog Description of the FUNC_FXT and Counter_STP Blocks 151
6.3.3 Verilog Description of the FUNC_PSI Block 154
6.3.4 Verilog Description of FUNC_S1 Block 155

6.3.5 Verilog Description of FSM_FR Block 156
6.3.6 Verilog Descriptions of SUM and Short Memory Blocks ... 157
6.3.7 Verilog Description of Block FRAC_MEM_TD_EQU 163
6.4 Synthesis of a PRNG Based on a Fractional-Order Memristive System with Time Delay 164
6.4.1 Time Series and Attractors Simulated within Matlab........ 164
6.4.2 PRNG from Chaotic Time Series by Applying mod256 167
6.4.3 Verilog Description of a PRNG and Simulation Results 167
6.5 PRNG Based on Fractional-Order Chaotic System: FPGA Implementation and Experimental Results 170
6.5.1 Experimental Time Series and Chaotic Attractor 171
6.5.2 Experimental Generation of Binary Strings.................. 172
6.6 Conclusion 172
References 174

7 Chaos-Based Encryption and Authentication of RGB Images 177
7.1 Introduction 177
7.2 Secure Communication System Using Chaos-Based PRNGs 178
7.3 Authentication of RGB Images.. 179
7.3.1 Hash Function Based on the Pseudo Dot Product 180
7.3.2 Operation of the Authentication Stage 181
7.4 Verilog Descriptions of the System for Authentication of Encrypted RGB Images.. 181
7.5 Verilog Simulation and FPGA Implementation Results............... 186
7.6 System for Text Authentication and Encryption....................... 188
7.7 Conclusions 189
References 191

Index 193

Chapter 1
Introduction to the Design and Implementation of PRNGs Based on Chaotic Maps and Systems

1.1 Introduction

Nowadays, it is well known and accepted that the word *chaos* is often taken as synonymous of the word disorder or unpredictability. The book titled *Philosophy of complex systems* [1] provides some philosophical comments related to the determinism chaos theory, paying emphasis that the word determinism can be associated to the dynamics of particular chaotic systems. However, the authors mention that determinism chaos of scientists has a very different meaning. For example, a geographical system, whose evolution obeys laws, is indeed chaotic. As the name of this theory indicates, the first law expresses the fact that a deterministic system may produce unpredictable long-term results. This law was discovered by the meteorologist E. Lorenz in 1963, whose main contribution was published in the work titled *Deterministic nonperiodic flow* [2], where it is claimed that finite systems of deterministic ordinary nonlinear differential equations may be designed to represent forced dissipative hydrodynamic flow. Currently, one can find a huge number of numerical methods to solve or approximate the solution of those differential equations, which can be identified with trajectories in phase space. E. Lorenz demonstrated that for those systems with bounded solutions, nonperiodic solutions are ordinarily unstable with respect to small modifications, so that slightly differing initial states can evolve into considerably different states. Another important contribution introduced in [2] is that systems with bounded solutions are shown to possess bounded numerical solutions.

The authors in [1] mention that indeed, the unpredictability of chaotic systems only seems to be evident over the long term. It is therefore possible to make short-term predictions. However, depending on the type of chaotic physical phenomena, the long term, as with the short term, depends upon the system studied. The long term in the context of the solar system encompasses billions of years, while the long term for the Icelandic Low does not, in fact, exceed a few days. The conclusion from the work introduced by E. Lorenz is that unpredictable behavior of chaotic systems

L. G. de la Fraga et al., *Random Number Generators*,
https://doi.org/10.1007/978-3-031-82865-2_1

has two origins: nonlinearity and sensitivity to initial conditions. Henceforth, one of the main characteristics associated to chaotic behavior is that the solution of a nonlinear system should be highly sensitive to initial conditions. This main property of any type of chaotic system be discrete as chaotic maps or continuous is the butterfly effect, very widely known and associated with its discoverer, E. Lorenz. The butterfly effect is defined as the simple fluttering of a butterfly's wings, which, highly amplified by a feedback effect, may change radically both the division of air masses and centers of action—depressions and anticyclones—across the globe.

Chaos theory describes the qualities of the point at which stability moves to instability or order moves to disorder. For example, unlike the behavior of a pendulum, which adheres to a predictable pattern, a chaotic system does not settle into a predictable pattern due to its nonlinear processes. Examples of chaotic systems include the mathematical models of chaotic oscillators implemented with electronic devices, as the ones described in Chaps. 2, 5, and 6, in this book. Henceforth, since the main property of chaotic systems is associated to the characteristic that they are highly sensitive to initial conditions, then this book shows the application of chaotic maps and chaotic systems of integer and fractional order [3], to the design of random number generators (RNGs). As there exist two main types of RNGs, namely, true and pseudo-RNGs, this book shows the design of PRNGs. For instance, the hardware synthesis of this type of chaos-based PRNGs requires the use of arithmetic and sequential digital blocks, which can be described under Verilog codes, as detailed in Chap. 3. However, the binary sequences that are generated must be validated to be random. This process can be performed through the verification of the binary strings as shown in Chap. 4, where one can see guidelines to perform TestU01 and NIST statistical tests for chaotic binary sequences.

The reference [1] states that advances in chaos theory and its mathematics are owed to physicist and mathematician Jules Henri Poincare (1854–1912), who used topological techniques to visualize mathematics. One can also find evidences that chaos mathematicians in the 1960s would map the trajectories, for example, of a simple pendulum. Nowadays, it is well known that the map or depiction can be described in the phase space, corresponding to the coordinates of the movement. For instance, a simple pendulum swing would have a two-dimensional phase space of velocity and angle. Once the movement is represented or mapped on the coordinates, a pattern appears. This pattern is called an *attractor*, and for chaotic systems the phase space among plotting their state variables is known as strange attractors. In recent decades, however, a diversity of systems have been studied that behave unpredictably despite their seeming simplicity and the fact that the forces involved are governed by well-understood physical laws. The common element in these systems is a very high degree of sensitivity to initial conditions and to the way in which they are set in motion. This is an important characteristic and justification of why chaotic maps and systems can be used to generate random binary strings, whose randomness can be verified by performing statistical tests, as TestU01 and NIST, which are described in Chap. 4.

Another important hot and new topic that combines chaotic behavior with intelligence has been presented in [4], where one can appreciate the potential of machine learning algorithms in addressing reliability and security by developing chaos-based encryption systems, which can be useful in secure communication applications, as in resource-constrained devices, such as wearable devices. This opens a new challenge on the development of modern applications in the search of machine learning techniques for understanding chaotic dynamical systems and improving signal synchronization. In the same direction, the open challenge that is related to the randomness of chaotic binary strings is the development of applications based on machine learning techniques to anticipate and mitigate attacks on chaos-based secure communication and transmission systems. These and other topics related to the generation of binary random sequences are a challenge that is more hard to accomplish when the chaotic maps or systems must be implemented on electronic devices, as shown in this book.

In [5], one can find examples of designing chaotic systems using integrated circuit technology, discrete devices, field-programmable analog arrays (FPAAs), microcontrollers, and field-programmable gate arrays (FPGAs). Due to the advantages for fast verification and prototyping, this book is focused on the synthesis of PRNGs based on chaotic maps and systems, using FPGAs. In this manner, the experimental FPGA implementations of chaotic maps and integer-order chaotic systems are described in Chap. 5. Chapter 6 shows the FPGA implementation of a PRNG when using a fractional-order chaotic system. The FPGA implementations of the PRNGs based on chaotic maps and systems guarantee the randomness of the binary strings, so that a slight change in the value of the initial condition of the state variables produces a vast impact on the values of the binary strings. An application for the authentication of color images encrypted by chaos-based PRNGs is presented in Chap. 7, where the transmission system is implemented on FPGAs.

1.2 True and Pseudo-Random Number Generators

A truly or true random number generator (TRNG) is a system that generates sequences of bits that are indistinguishable of noise with a uniform distribution. One example is given as follows: If a TRNG generates 16 bits every time step, then a TRNG of real numbers within the interval $[0, 1)$ can be created by dividing every sequence of 16 bits by the number $2^{16} - 1$, which in binary representation is composed by sixteen consecutive ones.

Basically, a TRNG can be generated by using a physical source of noise, such as thermal noise, the state of free oscillators, or chaos [6–8].

A pseudo-random number generator (PRNG) is a deterministic system, which will always generate the same sequence of binary numbers given the same *seed*. The generated sequence is called pseudo-random, because it seems statistically random—the sequence cannot be distinguished of a uniform distribution noise—but the same sequence can be generated exactly by using the same seed. In general, this seed can be an integer number, or the initial state values for a chaotic system [9–11]. In this manner, as a PRNG is a deterministic system, then it can be created with the use of a computer. In our point of view, a PRNG is a software or hardware system that uses chaos as a source of entropy for the generation of the pseudo-random binary sequences. It is for this reason that this book is devoted to provide a very deep study on how to design PRNGs based on chaotic maps or integer- and fractional-order chaotic systems.

The authors in [12] mention that the typical TRNG structure can be divided into five modules:

1. An analog random signal is obtained from the entropy source.
2. A module that performs sampling and the process of quantifying the random signal.
3. A module for the analog-to-digital conversion of the analog signal to output the random number sequence.
4. The sequence obtained at this time does not necessarily satisfy the uniform distribution, and it needs to be processed.
5. A module to validate the binary strings through a random number test suite.

A very effective technique to process a binary sequence is the *bit counting redundancy reduction technique*, which has been used in [13]. The process can be summarized as follows: The original binary sequence is divided in blocks of 5 bits, and then a new bit for a new sequence is generated by applying the XOR logic operation on the 5 bits of each block. This technique was also applied in [9] to create PRNGs from different types of one-dimensional (1D) chaotic maps.

One can agree with the concept that a binary sequence can be considered random, if it is computationally difficult for an observer with reasonable computational resources at their disposal, to predict it. Or in other words, the binary sequence cannot be distinguished from noise.

In the side of applications based on chaotic maps or systems, a mathematical proof does not exist where the generated binary sequences are random. Instead, statistical tests are used to prove if a very long binary sequence is random or not. Among the currently available statistical tests, this book revises two of them that are more used, namely, NIST and TestU01 tests [14, 15]. The former comes from the National Institute of Standards and Technology https://www.nist.gov/director/pao/nist-general-information, which is an agency of the United States Department of Commerce whose mission is to promote American innovation and industrial competitiveness. The second statistical test suite is the TestU01, which is a software library, implemented in the ANSI C language, that offers a collection of utilities for the empirical randomness testing of RNGs https://simul.iro.umontreal.ca/testu01/tu01.html. These statistical test suites are described in detail in Chap. 4, taking, as case study, binary sequences generated from chaotic systems.

A PRNG is essential to develop applications in engineering areas such as to improve the behavior of optimization algorithms with heuristics. Some of those algorithms are heuristics well known from a couple of decades, such as genetic algorithm, differential evolution, and particle swarm optimization, among much other evolutionary and swarm algorithms, which perform a guided (intelligent) search. In fact, the intelligent search is a much better approach compared to the brute force or random search approaches.

There is an approach in statistics called Monte Carlo, where the acknowledge is acquired randomly (or pseudo-randomly).

As shown in [4], PRNGs can also be used to improve machine learning applications. Indeed, in those applications a PRNG is essential. Basically, PRNGs are used to select randomly the element to train and to test a machine learning algorithm.

It can be appreciated that there are many more applications that need RNGs. Some of the most known are, for example, [6]: entertainment, where lotteries and gambling machines are all based on the use of random numbers, video games that use random numbers to influence intelligence heuristics or to vary game play, music and graphics composition by interweaving content with random bits, complex scientific and financial models that use random numbers for simulation, artificial intelligence applications that require random data to determine classification accuracy or neural network behavior, and software manufacturers that use random data to test programs and algorithms to detect bugs, equation-solving, cryptography, digital signatures, protected communication protocols, and so on.

1.3 Chaos and Cryptography

The two scientific areas of chaos and cryptography lie in the fact that the systems used in cryptography work on a finite set, while those applied in chaos have meaning only on a continuum [16]. This has been a big problem in chaos systems, where, as an example, it is not possible to define the exact size of the keys, even if the key is used with parameter values of real numbers. On the other hand and as it will be explained in this book, if a PRNG is implemented in fixed-point arithmetic [9–11], then it is possible to define the exact size of the key.

The authors in [16] present a good description of the three most common cryptographic objects (called also primitives):

1. Block-encryption algorithms (private-key algorithms)
2. Pseudo-random number generators (additive stream ciphers)
3. Cryptographic hash functions

A PRNG is the basic primitive to define block-encryption algorithms, cryptographic hash functions, and even keyed hash functions [17]. For instance, a private-key algorithm can be built with a chaos-based PRNG if the two involved persons in a communication channel share the key, being the key the initial state of the map or chaotic oscillator, and then the message is XORed with the sequence produced by the PRNG; the receiver can decode the message producing the pseudo-sequence of bytes with the PRNG and the same key and XOR the pseudo-random sequence with the coded message to obtain the text (decrypted) message.

A hash is a small quantity of bytes that can be used as a digital signature for a document: If the document is sent over the Internet, the calculated hash on the received node must be equal to the original or sent hash. However, if both the send and received hashes are not equal, then it can be concluded that the document has been compromised during its transmission. For this reason, it is commonly accepted that a keyed hash function [17], as its name suggest, can add a key to calculate the hash, and then it can be used for both, to check the integrity and authenticate a message.

At the present time, the notion of cryptographic security has no counterpart in chaos theory, and the cryptographic security of a chaos-derived encryption algorithm can be checked only by means of crypto-tools.

1.4 Generation of Pseudo-Random Numbers with Available Software

The goal of this book is the design of PRNGs based on chaotic maps and integer- and fractional-order chaotic systems. However, this section is devoted to describe how a pseudo-random binary sequence can be generated with publicly available cryptography software. To do this task, one can apply software tools, such as a software package called *OpenSSL* and the software that is a family of lightweight cryptographic algorithms called *ASCOM*. As shown in Chap. 4, the generated sequences can be analyzed with the NIST and TestU01 statistical suite tests, just as examples of sequences that already are known to be cryptographically secure.

The software package OpenSSL [18] is a cryptography toolkit implementing the Secure Sockets Layer (SSL) and Transport Layer Security (TLS) network protocols and related cryptography standards required by them. OpenSSL is the layer of software that allows secure communications between or among computers or networks, tablets, or smartphones, over the Internet. Using OpenSSL guarantees that online buy transactions will be completed, and that credit card numbers will be not stolen while those buys are made.

OpenSSL is available for any operating system. In fact, it is already included in any device that is connected to the Internet. OpenSSL can manage keys and encrypt and decrypt messages with different cyphers, among other abilities.

An advantage of using OpenSSL is that one file with 100 millions of pseudo-random bits can be generated with the command

```
openssl rand -out a.bin 12500000
```

where the output file is `a.bin`, and it gives the number of bytes instead the number of bits to be generated.

Executing the NIST and TestU01 statistical test suites on the `a.bin` file, they give the results shown in Table 1.1, where it can be appreciated that all statistical tests passed. Some considerations on the statistical tests are as follows: For 100 sequences, a test pass if the proportion value is greater than 96, with a p-value

Table 1.1 Results of applying the three TestU01 tests and fifteen NIST statistical tests on 100 sequences of 10^6 bits generated with the OpenSSL software

	Test name	p-value	Proportion
1	Rabbit	All 40 tests passed	–
2	Alphabit	All 17 tests passed	–
3	Block Alphabit	All 6 repetitions of Alphabit tests passed	–
4	Frequency	99	0.048716
5	BlockFrequency	99	0.383827
6	CumulativeSums	99	0.734931
7	Runs	100	0.867692
8	LongestRun	100	0.574903
9	Rank	100	0.045675
10	FFT	99	0.719747
11	NonOverlappingTemplate	99	0.489748
12	OverlappingTemplate	100	0.978072
13	Universal	99	0.779188
14	ApproximateEntropy	100	0.249284
15	RandomExcursions	99	0.449323
16	RandomExcursionsVariant	100	0.403272
17	LinearComplexity	97	0.514124
18	Serial	100	0.030184

greater than or equal to 0.01, and also the statistics of the p-values must have a uniform distribution. Therefore, if the p-value given in Table 1.1 is equal or greater than 0.0001, it can be concluded that the test pass [14, Chap. 4.2].

The software that is a family of lightweight cryptographic algorithms called *ASCOM* is also a family of authenticated encryption and hashing algorithms designed to be lightweight and easy to implement, even with added countermeasures against side-channel attacks. In this case, a side-channel attack can be implemented with the working algorithm in hardware, where the power consumption can reveal the key used. ASCOM has been selected as a new standard for lightweight cryptography in the NIST Lightweight Cryptography competition (2019–2023) [19, 20].

The publicly available software of ASCOM [19] uses the interface *ECRYPT Benchmarking of Cryptographic Systems* (eBACS). As an example, the encryption function in C program language can be coded as follows:

```
int crypto_aead_encrypt(
    unsigned char *cmsg, unsigned long long *clen,
    const unsigned char *m, unsigned long long mlen,
    const unsigned char *ad, unsigned long long adlen,
    const unsigned char *nsec, const unsigned char *nonce,
    const unsigned char *key
);
```

where the instruction `unsigned char *cmsg` is associated to the encrypted message and the instruction `const unsigned char *m` associated to the text message.

Similar to the generation of random binary strings by executing the software package OpenSSL [18], in this case the binary strings are generated by applying the software ASCOM. The example consists of generating a block of 16 bytes that is used to encrypt a message of also 16 bytes, but consisting of only zeros. The nonce values must be different to produce different encrypted blocks. To perform statistical tests, 100 binary sequences of 10^6 bits each one were generated by executing the C code given in Listing 1.1. Afterwards, the TestU01 and NIST statistical test suites were applied to the 100 binary sequences, in which statistical results are given in Table 1.2.

Listing 1.1 C code to generate random bytes with ASCOM

```c
#include <stdio.h>
#include <stdlib.h>
#include <string.h>
#include "api.h"
#include "crypto_aead.h"

/** This function calculates the nonce,
  Converts 'i' value to a string  **/
void calculate_nonce( unsigned char *v, int i )
{
  int k, c, r;

  k = 0;
  while( k < 16) {
    c = i/256;
    r = i - 256*c;

    v[k] = r;
    i = c;
    k++;
    if( i < 256 ) {
      v[k] = i;
      break;
    }
  }
}

int main( )
{
  #define NSIZE  64
  FILE *fp;
  /** 16 bytes the size of the key and the nonce **/
  unsigned char key[CRYPTO_KEYBYTES] =
      {0,1,2,3,4,5,6,7,8,9,0,1,2,3,4,5};
  unsigned char nonce[CRYPTO_NPUBBYTES] = { 0x00 };
  unsigned char msg[NSIZE] = { 0x00 };
  unsigned char* ad = NULL;
  unsigned char ct[NSIZE+CRYPTO_ABYTES]; // 64+16 bytes
```

```
  unsigned long long mlen, adlen;
  unsigned long long clen, mlen2;
  int i, func_ret;

  mlen = NSIZE;
  adlen = 0;

  if( (fp=fopen("a.bin", "w" )) == NULL ) {
    fprintf( stderr, "ERROR: can't open output file\n" );
    exit(1);
  }
  // blocks of size 64*8 =  512 bits **/
  for( i = 0; i<195314; i++ ) {
    calculate_nonce( nonce,  i );
    if ((func_ret = crypto_aead_encrypt(ct, &clen, msg, mlen, ad,
         adlen, NULL,
                                    nonce, key)) != 0) {
      fprintf(fp, "crypto_aead_encrypt returned <%d>\n", func_ret
         );
      exit(2);
    }
    if ( fwrite ( ct, 1, NSIZE, fp) != NSIZE ) {
      fprintf ( stderr, "ERROR: writing file: %s\n", "a.bin" );
      exit(3);
    }

  }

  fclose( fp );
  return 0;
}
```

The goal of this book is also oriented to infer that the future of cryptographic chaotic research must be more focused to design efficient software and hardware that can offer better security services.

1.5 Hardware Implementation of Pseudo-Random Number Generators

In the recent days, one can find a huge number of topologies for the design of chaotic maps and systems using electronic devices. They range from using discrete devices that are commercially available, such as amplifiers and microcontrollers, and also those using modern complementary-metal-oxide-semiconductor (CMOS) integrated circuit technology. This book is devoted to show the synthesis of PRNGs based on chaotic maps and integer- and fractional-order chaotic systems by using FPGAs. However, one can also find PRNGs implemented on field-programmable analog arrays (FPAAs). The following paragraphs briefly describe some of the most recent electronic implementations of chaotic systems and PRNGs.

Table 1.2 Results of applying the three TestU01 tests and fifteen NIST statistical tests on 100 sequences of 10^6 bits generated with ASCOM with the function shown in Listing 1.1

	Test name	p-value	Proportion
1	Rabbit	All 40 tests passed	–
2	Alphabit	All 17 tests passed	–
3	Block Alphabit	All 6 repetitions of Alphabit tests passed	–
4	Frequency	100	0.657933
5	BlockFrequency	98	0.798139
6	CumulativeSums	100	0.742930
7	Runs	100	0.935716
8	LongestRun	100	0.514124
9	Rank	100	0.798139
10	FFT	100	0.867692
11	NonOverlappingTemplate	99	0.481414
12	OverlappingTemplate	97	0.262249
13	Universal	100	0.779188
14	ApproximateEntropy	99	0.759756
15	RandomExcursions	99	0.571658
16	RandomExcursionsVariant	99	0.305836
17	LinearComplexity	98	0.574903
18	Serial	99	0.167346

The work given in [21] shows the CMOS design of Lorenz systems using operational transconductance amplifiers (OTAs) designed with an integrated circuit technology of 180 nm. The authors show the correct design of a biquadratic filter to guarantee chaotic behavior. The chaotic time series of this CMOS Lorenz system can be postprocessed to generate binary strings, as by simple using comparators, and then it is possible to design a PRNG. The CMOS integrated circuit technology of 180 nm has also been used in [22], to design a grounded meminductor emulator architecture based on only MOSFETs and two capacitors. In [23], the authors claim that a challenge of physically realizing memelements is an open issue that encourages the development of newer concepts. In this manner, the authors proposed a multimem-element architecture by changing one impedance of the circuit as resistor, capacitor, or inductor. The proposed multimem-element was designed with CMOS technology, and it has the capability of emulating all the three memelements, namely, memristor, memcapacitor, and meminductor. The architecture is also suitable to add a comparator to generate binary strings. Other recent application of a meminductor designed with CMOS technology is given in [24], to design a chaotic oscillator. The memristor has also been designed with CMOS integrated circuits, as shown in [25–27], which architectures can be exploited to design PRNGs.

From five decades ago, several chaotic systems have been implemented using discrete devices, which are commercially available, such as operational amplifiers, current-feedback operational amplifiers, current conveyors, and so on. The work given in [28] presents the application of a modified differential voltage current conveyor transconductance amplifier, to design a meminductor emulator, which can be used to design a chaotic oscillator. A similar modern circuit element is presented in [29], for the design of Chua's diode by using a current-controlled current conveyor transconductance amplifier, with an active MOS transistor for electronically tunable feature. In this manner, Chua's circuit is designed using the proposed Chua's diode, generating a hidden chaotic attractor with one unstable node equilibrium and two symmetric stable node equilibrium points.

In the side of digital electronics, one can find a huge number of implementations of chaotic systems using microcontrollers. The authors in [30] implemented a memristive state-controlled cellular neural network on a microcontroller to generate multi-double-scroll attractors in one direction or grid multi-double-scroll attractors. The microcontroller was also applied to design chaotic maps, as shown in [31], for the implementation of a chaotic map with two-dimensional offset boosting, allowing controllability of amplitude control. Some memristor-based chaotic systems have been implemented by using microcontrollers. The authors in [32] implemented a fractional-order memristor-coupled Hindmarsh-Rose neuron model, for which when external stimulus is applied to a neuron, the system produces anti-monotonicity and bursting behavior. The microcontrollers have been applied to design chaotic systems and PRNGs in the works [33–36]. In particular, some authors presented implementations of chaotic systems by using the STM-32 microcontroller, as shown in [37, 38].

The use of programmable devices, as the FPAA, has also the potential of implementing chaotic systems, as shown in [39, 40]. This programmable device has also solved open problems as the one presented in [41], where the authors claim that the ability to obtain offset boosting independently in two dimensions within a chaotic system can meet the demands of multi-signal conditioning. In [42], the authors present the implementation of an image cryptography communication system using FPAA-based multi-scroll chaotic systems. In [43], the FPAA is used to implement a real-time neural identification system using a recurrent wavelet first-order neural network in a chaotic system. Recently spiking behavior is a hot topic and the work in [44] presents the implementation of various chaotic spiking oscillators based on FPAAs. In the case of fractional-order chaotic systems, the following works show the usefulness of using FPAAs for their implementations [45–49].

In the case of FPGAs, they have been widely applied in the design of TRNGs and PRNGs. The work in [50] shows the implementation of a gigabit TRNG with a novel lightweight post-processing method for cryptographic applications. The authors in [51] introduce a perspective of using frequency mixing as entropy in TRNGs for portable hardware cybersecurity IP. The ring oscillator-based physically unclonable function can be implemented into an FPGA for high-throughput Internet of Things (IoT) applications, as shown in [52]. In [53], the authors mention that modern technologies of truly flexible integrated circuits are usually based on large single-type transistors. If considered as a platform for IoT devices, they would require dedicated cryptographic solutions that could fit their limitations. In such a way the authors introduced the implementation of an efficient and fast-start TRNG developed with a metal-oxide thin-film transistor (TFT) technology on a flexible substrate.

The design of PRNGs based on chaotic maps and integer- and fractional-order chaotic systems is the goal of this book. Some FPGA-based implementations of PRNGs can be summarized as follows: The work [54] presents the design of a high-speed PRNG based on an n-D nondegenerate chaotic system. The authors in [55] present a 4D Hamiltonian conservative chaotic system with line equilibrium and grid-scroll chaotic flows. A 4D trigonometric-based memristor hyperchaotic map to ultrafast PRNG is presented in [56]. An artificial neural network chaotic PRNG and simple encryption on FPGA are presented in [57]. Other recent FPGA implementations of PRNGs based on chaotic systems can be read in [58–63].

1.6 Conclusion

This chapter summarized the general concepts on chaos theory and emphasized the need of generating binary random sequences. It was mentioned that there are two types of RNG, namely, TRNG and PRNG. Both types can be used to develop applications in cryptography, where a couple of examples for the generation of random binary strings by available software packages were shown. Besides, the goal of this book is the FPGA implementation of PRNGs.

The following chapters provide details on the following issues: Chap. 2 shows the mathematical modeling of integer- and fractional-order chaotic systems and describes their simulation by numerical methods. The discretized equations of the chaotic systems can then be described by digital blocks, whose Verilog codes are detailed in Chap. 3. The hardware descriptions are performed by using fixed-point formats, so that the generated binary sequences must be tested to verify their randomness. This is described in Chap. 4, where one can understand the way TestU01 and NIST statistical tests can be performed. The FPGA implementation and the experimental results of PRNGs based on chaotic maps and chaotic systems are shown in Chap. 5, while Chap. 6 shows the FPGA synthesis of PRNGs based on a fractional-order chaotic system. Finally, Chap. 7 shows an application of a PRNG for the authentication of encrypted RGB images.

References

1. D.M. Gabbay, P. Thagard, J. Woods, C.A. Hooker, *Philosophy of Complex Systems*, vol. 10 (Elsevier, Amsterdam, 2011)
2. E.N. Lorenz, Deterministic nonperiodic flow. J. Atmos. Sci. **20**(2), 130–141 (1963)
3. I. Petráš, I. Petráš, Fractional-order chaotic systems, in *Fractional-Order Nonlinear Systems: Modeling, Analysis and Simulation* (Springer, Berlin, 2011), pp. 103–184

4. J. Hwang, G. Kale, P.P. Patel, R. Vishwakarma, M. Aliasgari, A. Hedayatipour, A. Rezaei, H. Sayadi, Machine learning in chaos-based encryption: theory, implementations, and applications. IEEE Access **11**, 125749–125767 (2023)
5. E. Tlelo-Cuautle, A.D. Pano-Azucena, O. Guillén-Fernández, A. Silva-Juárez, *Analog/Digital Implementation of Fractional Order Chaotic Circuits and Applications* (Springer, Berlin, 2020)
6. B. Jun, P. Kocher, The intel random numbero generator. Cryptography Research Inc. white paper, San Francisco (1999)
7. M. Stipčević, Ç.K. Koç, *True Random Number Generators* (Springer International Publishing, Cham, 2014), pp. 275–315
8. B. Karakaya, A. Gülten, M. Frasca, A true random bit generator based on a memristive chaotic circuit: analysis, design and fpga implementation. Chaos Solit Fract **119**, 143–149 (2019)
9. L.G. de la Fraga, E. Torres-Perez, E. Tlelo-Cuautle, C. Mancillas-López, Hardware implementation of pseudorandom number generators based on chaotic maps. Nonlinear Dyn. **90**(3), 1661–1670 (2017)
10. L.G. de la Fraga, B. Ovilla-Martínez, A chaotic prng tested with the heuristic differential evolution. Integr. VLSI J. **90**, 22–26 (2023)
11. L.G. de la Fraga, B. Ovilla-Martínez, Generating pseudo-random numbers with a brownian system. Integration **96**, 102135 (2024)
12. F. Yu, L. Li, Q. Tang, S. Cai, Y. Song, Q.N. Xu, A survey on true random number generators based on chaos. Discret. Dyn. Nat. Soc. **2019**(1), 2545123 (2019)
13. T. Stojanovski, J. Pihl, L. Kocarev, Chaos-based random number generators. Part ii: practical realization. IEEE Trans. Circuits Syst. Fundam. Theory Appl. **48**(3), 382–385 (2001)
14. A. Rukhin, J. Soto, J. Nechvatal, M. Smid, E. Barker, S. Leigh, M. Levenson, M. Vangel, D. Banks, A. Heckert, J. Dray, S. Vo, A statistical test suite for the validation of random number generators and pseudo random number generators for cryptographic applications, April 27, 2010. SP800-22rev1a.pdf file, last date checked: Jan 31, 2024. Online at http://csrc.nist.gov/groups/ST/toolkit/rng/documentation_software.html
15. L. Ecuyer, R. Simard, TestU01: A C library for empirical testing of random number generators. ACM Trans. Math. Softw. **33**(22), (2007). http://simul.iro.umontreal.ca/testu01/tu01.html
16. L. Kocarev, Chaos-based cryptography: a brief overview. IEEE Circuits Syst. Mag. **1**(3), 6–21 (2001)
17. L.G. de la Fraga, C. Mancillas-López, E. Tlelo-Cuautle, Designing an authenticated hash function with a 2d chaotic map. Nonlinear Dyn. **104**, 4569–4580 (2021)
18. Openssl Library, Last time checked: Jul 24, 2024. https://openssl.org
19. Ascon, lightweight authenticated encryption & hashing, Last time checked Jul 19th, 2024. https://ascon.iaik.tugraz.at
20. B. Ovilla-Martínez, C. Mancillas-López, A.F. Martínez-Herrera, J. Bernal-Gutiérrez, FPGA implementation of some second round NIST lightweight cryptography candidates. Electronics **9**, 1940 (2020)
21. E. Juarez-Mendoza, F.A. del Angel-Diaz, A. Diaz-Sanchez, E. Tlelo-Cuautle, CMOS design of chaotic systems using biquadratic OTA-C filters. J. Low Power Electr. Appl. **14**(1), 14 (2024)
22. P. Kumar, A.M. Bhat, P.K. Sharma, R.K. Ranjan, A novel and compact MOSFET-C only based grounded meminductor emulator and its application. AEU-Int. J. Electro. Commun. **183**, 155378 (2024)
23. N. Raj, R.K. Ranjan, Emulation of novel floating and tunable multimem-elements circuit and its application. AEU-Int. J. Electro. Commun. **177**, 155215 (2024)
24. S.K. Rai, B. Aggarwal, V. Singroha, Meminductor emulators using off-the-shelf active blocks with application in chaotic oscillator. Integrat. VLSI J. **96**, 102165 (2024)
25. K.S. Seetala, W. Clower, M. Hartmann, S. Zivanovic, Physical implementation of cobalt ferrite memristor in chua's circuit for chaotic encryption. Microelectro. Eng. **290**, 112194 (2024)
26. Y. Zhao, F. Parastesh, S. He, L. Fu, S. Jafari, Integrated circuit design of a discrete memristive chaotic system optimized by the fixed-point specific processor with acceleration instructions. Nonlinear Dyn. **112**(12), 10451–10464 (2024)

27. P. Kumar, R.K. Ranjan, S.-M. Kang, A memristor emulation in 180-nm cmos process for spiking signal generation and chaos application. IEEE Trans. Circuits Syst. I-Regul. Pap. **71**(4), 1757–1770 (2024)
28. R. Das, S.K. Rai, B. Aggarwal, A floating meminductor emulator using modified differential voltage current conveyor transconductance amplifier and its application. Analog Integr. Circuits Signal Process. **119**(3), 475–496 (2024)
29. V. Bhatt, A. Ranjan, M. Joshi, Ccccta-based chua's circuit for chaotic oscillation. Circuits Syst. Signal Process. **43**(4), 2051–2072 (2024)
30. H. Jin, Z. Li, Generating grid double-scroll attractors from magnetized SC-CNN and its application in image encryption. Chaos Solit. Fract. **186**, 115219 (2024)
31. Z. Wang, C. Li, Y. Li, S. Liu, A. Akgul, A chaotic map with two-dimensional offset boosting. Chaos **34**(6), 063130 (2024)
32. D. Ding, S. Chen, H. Zhang, Z. Yang, F. Jin, X. Liu, Firing pattern transition of fractional-order memristor-coupled hindmarsh-rose neurons model and its medical image encryption for region of interest. Nonlinear Dyn. **112**(12), 10529–10554 (2024)
33. G. Sivaganesh, K. Srinivasan, T. Fozin Fonzin, P.D. Kamdem Kuate, I.R. Mohamed, Analytical studies on third-order chaotic systems with sprott type nonlinearities and their microcontroller implementation. Phys. Scr. **99**(4), 045246 (2024)
34. R. Mendez-Ramirez, A. Arellano-Delgado, M. Murillo-Escobar, C. Cruz-Hernandez, Digital synchronization of the MACM chaotic system by using PIC24-microcontrollers and the SPI-protocol. Integr. VLSI J. **96**, 102158 (2024)
35. B. Nail, M.A. Atoussi, S. Saadi, I.E. Tibermacine, C. Napoli, Real-time synchronisation of multiple fractional-order chaotic systems: an application study in secure communication. Fractal Fract. **8**(2), 104 (2024)
36. D. Murillo-Escobar, K. Vega-Perez, M.A. Murillo-Escobar, A. Arellano-Delgado, R.M. Lopez-Gutierrez, Comparison of two new chaos-based pseudorandom number generators implemented in microcontroller. Integr. VLSI J. **96**, 102130 (2024)
37. F. Li, L. Bai, Z. Chen, B. Bao, Scroll-growth and scroll-control attractors in memristive bi-neuron Hopfield neural network. IEEE Trans. Circuits Syst. II-Express Briefs **71**(4), 2354–2358 (2024)
38. N. Charalampidis, A. Iatropoulos, C. Volos, Chaos based speech encryption using microcontroller. Integr. VLSI J. **95**, 102128 (2024)
39. N. Korkmaz, İ.E. Saçu, An efficient design procedure to implement the fractional-order chaotic jerk systems with the programmable analog platform. Chaos Theory Appl. **3**(2), 59–66 (2021).
40. X. Li, A. Beal, R. Dean, E. Perkins, Chaos in a pendulum adaptive frequency oscillator circuit experiment. Chaos Theory Appl. **5**(1), 11–19 (2023)
41. W. Liu, C. Li, Y. Gao, X. Zhang, Y. Xu, J. Xu, Simplified chaotic oscillators with two-dimensional offset boosting. Eur. Phys. J. Plus **139**(7), 596 (2024)
42. K. Karawanich, J. Chimnoy, F. Khateb, M. Marwan, P. Prommee, Image cryptography communication using FPAA-based multi-scroll chaotic system. Nonlinear Dyn. **2024**, 4951 (2024)
43. D.A. Magallon-Garcia, J.H. Garcia-Lopez, G. Huerta-Cuellar, R. Jaimes-Reategui, I.A. Diaz-Diaz, L.J. Ontanon-Garcia, Real-time neural identification using a recurrent wavelet first-order neural network of a chaotic system implemented in an FPAA. Integr. VLSI J. **96**, 102134 (2024)
44. Y. Matsuoka, Implementation of various chaotic spiking oscillators based on field programmable analog array. IEICE Trans. Fundament. Electr. Commun. Comput. Sci. **E106A**(11), 1432–1435 (2023)
45. K. Iskakova, M.M. Alam, S. Ahmad, S. Saifullah, A. Akguel, G. Yilmaz, Dynamical study of a novel 4d hyperchaotic system: an integer and fractional order analysis. Math. Comput. Simulat. **208**, 219–245 (2023)
46. J. Hasler, A programmable on-chip hopf bifurcation circuit. IEEE Trans. Circuits Syst. I-Regul. Pap. **69**(12), 4958–4968 (2022)

47. K. Altun, Multi-scroll attractors with hyperchaotic behavior using fractional-order systems. J. Circuits Syst. Comput. **31**(05), 2250085 (2022)
48. K. Altun, FPAA implementations of fractional-order chaotic systems. J. Circuits Syst. Comput. **30**(15), 2150271 (2021)
49. A. Silva-Juarez, E. Tlelo-Cuautle, L.G. de la Fraga, R. Li, FPAA-based implementation of fractional-order chaotic oscillators using first-order active filter blocks. J. Adv. Res. **25**(SI), 77–85 (2020)
50. A.M. Garipcan, E. Erdem, A gigabit trng with novel lightweight post-processing method for cryptographic applications. Eur. Phys. J. Plus **137**(4), 493 (2022)
51. X. Wei, L. Xiu, Y. Cai, A perspective of using frequency-mixing as entropy in random number generation for portable hardware cybersecurity IP. IEEE Trans. Inf. Forens. Secur. **19**, 320–333 (2024)
52. A. Magyari, Y. Chen, Integrating lorenz hyperchaotic encryption with ring oscillator physically unclonable functions (ro-pufs) for high-throughput internet of things (IoT) applications. Electronics **12**(24), 4929 (2023)
53. K. Golofit, P.Z. Wieczorek, M. Pilarz, A chaos-metastability TRNG for natively flexible IGZO circuits. AEU-Int. J. Electr. Commun. **170**, 154835 (2023)
54. Y. Luo, C. Fan, C. Xu, X. Li, Design and FPGA implementation of a high-speed PRNG based on an N-D non-degenerate chaotic system. CHAOS Solit. Fract. **183**, 114951 (2024)
55. M. Ji'e, H. Peng, S. Duan, L. Wang, F. Zhang, D. Yan, Design and FPGA implementation of grid-scroll hamiltonian conservative chaotic flows with a line equilibrium. IEEE Trans. Very Large Scale Integr. (VLSI) Syst. **32**(4), 658–668 (2024)
56. B. Xu, X. Luo, Y. Wang, L. Bai, K. Chen, J. Zhao, A 4d trigonometric-based memristor hyperchaotic map to ultra-fast prng. IEEE Trans. Ind. Inf. **20**(6), 8673–8683 (2024)
57. B.K. Sharobim, M.H. Yacoub, W.S. Sayed, A.G. Radwan, L.A. Said, Artificial neural network chaotic PRNG and simple encryption on FPGA. Eng. Appl. Artif. Intell. **126**(B), 106888 (2023)
58. J. Tang, Z. Zhang, P. Chen, Z. Huang, T. Huang, A simple chaotic model with complex chaotic behaviors and its hardware implementation. IEEE Trans. Circuits Syst. I-Regul. Pap. **70**(9), 3676–3688 (2023)
59. Y. Li, M. Yuan, Z. Chen, Z. Chen, Coexistence and ergodicity in a variant Nose-Hoover oscillator and its FPGA implementation. Nonlinear Dyn. **111**(11), 10583–10599 (2023)
60. V.B.E. Mebenga, V.R. Kopparthi, H.D. Nzeuga, J.S.A.E. Fouda, G.M.D. Dagoumguei, G.B. Bitjoka, P. Rangababu, S.L. Sabat, An 8-bit integer true periodic orbit PRNG based on delayed Arnold's cat map. AEU-Int. J. Electr. Commun. **162**, 154575 (2023)
61. S. Li, Y. Liu, F. Ren, Z. Yang, Design of a high throughput pseudorandom number generator based on discrete hyper-chaotic system. IEEE Trans. Circuits Syst. II-Express Briefs **70**(2), 806–810 (2023)
62. F. Zhang, J. Tang, Z. Zhang, Z. Huang, T. Huang, An improved absolute-cosine chaotification model and its simple application in prng. IEEE Access **11**, 59346–59356 (2023)
63. M. Gafsi, R. Amdouni, M.A. Hajjaji, A. Mtibaa, E.-B. Bourennane, Hardware implementation of a strong pseudorandom number generator based block-cipher system for color image encryption and decryption. Int. J. Circuit Theory Appl. **51**(1), 410–436 (2023)

Chapter 2
Numerical Methods to Approximate Integer-/Fractional-Order Chaotic Systems

2.1 Introduction

Nonlinear systems, like the ones generating chaotic behavior, are an active field for research that is quite interesting to engineers, physicists, and mathematicians, because most real physical systems are inherently nonlinear in nature. Nonlinear systems can be modeled by nonlinear equations, but yet they are quite difficult to be solved by analytical methods, so that one applies numerical methods to approximate the solutions, which in some cases give rise to interesting phenomena such as bifurcation and chaos [1, 2].

It is well known that the word chaos can be associated with unpredictability or disorder and that the dynamical systems presenting chaotic behavior have super-sensibility to initial conditions, which makes them stochastic. Chaotic systems are a type of nonlinear systems in which the variables to be solved cannot be written as a linear combination of independent components, but in most cases their models can be described by ordinary differential equations (ODEs) that can include polynomial, trigonometric, or nonlinear functions. For instance, nonlinear systems including piecewise linear (PWL) functions can exhibit completely an unpredictable behavior, the so-called deterministic chaos. In this manner, chaotic behavior can also be defined as a long-term aperiodic behavior of a deterministic system that exhibits dependence and becomes extremely sensitive to initial conditions. Henceforth, an autonomous nonlinear system that is described by at least three (3D) ODEs can be considered as chaotic if it possesses at least one positive Lyapunov exponent. On this regard, there exists one positive Lyapunov exponent for each ODE, and thus, 3D systems can generate chaotic behavior having one positive, one zero, and one negative Lyapunov exponent. However, high-dimensional systems, as the systems with more than three ODEs, can have one or more positive, one or more zero, and one or more negative Lyapunov exponents. Therefore, a nonlinear system having

L. G. de la Fraga et al., *Random Number Generators*,
https://doi.org/10.1007/978-3-031-82865-2_2

more than one positive Lyapunov exponent is actually called hyperchaotic system. Besides, since chaotic and hyperchaotic systems are well defined by nonlinear ODEs, and although their behavior seems to be irregular or random, its nature is fully deterministic, that is, there is a set of ODEs that define their dynamics and from them one can know with good certainty and precision, the data sequence being generated. In summary, a chaotic system possesses the following characteristics: High sensitivity to initial conditions; deterministic characteristic because their parameters are known; as time increases ($t \rightarrow \infty$) its behavior is difficult to distinguish, and the trajectories of the state variables do not tend to a fixed point, periodic, or quasi-periodic orbits; the strange attractors have a fractal dimension; and it has at least one positive Lyapunov exponent.

Performing a search on scientific databases for the keywords related to chaotic systems, one can identify that in a wide range of engineering research areas, chaotic circuits and systems are taken into account to introduce new mathematical models and improved electronic implementations to enhance chaos-based applications, such as secure communications [3], Internet of Things (IoT) [4], neuromorphic [5, 6], robotics [7, 8], authentication [9], and encryption [10]. Not only chaotic systems of integer ODEs are a hot topic, but also those of fractional order, which exploit fractional calculus and allow implementations with analog and digital electronics [11]. Integer- and fractional-order chaotic systems can generate self-excited and hidden attractors. In the former case, the self-excited attractor has a basin of attraction that is excited from unstable equilibrium points, as the classical nonlinear systems such as Lorenz [12], Chen [13], and Rössler systems [14]. The other case is hidden attractors, which cannot be localized straightforwardly applying manual calculations, so that one must develop specific computational procedures to identify them because the evaluation of their equilibrium points is difficult and they do not help in their localization.

2.2 Numerical Methods for Integer-/Fractional-Order Chaotic Systems

Chaos is a nonlinear and unpredictable behavior that for autonomous systems, it can be modeled as an initial value problem (IVP) of the form $\dot{x} = f(x)$, which requires initial condition values $x(0)$ to be solved. In continuous-time modeling, the minimum number of ODEs for autonomous chaotic oscillators is three, and henceforth, dynamical systems modeled by four or more ODEs can generate hyperchaotic behavior. The solution to an IVP can be performed by applying numerical methods [15], in which the time-step h value is quite important to generate chaotic time series that can be used to estimate Lyapunov exponents, as already shown in the seminal work [16].

2.2.1 *Numerical Methods for Integer-Order Chaotic Systems*

In [11], one can find the application of numerical methods to solve an IVP associated to integer- and fractional-order chaotic systems that are implemented with electronic devices. In the case of chaotic systems of integer order, the IVP can be solved by applying one-step and multistep methods. Basically, the former methods use one past value evaluated at iteration x_i to compute x_{i+1}, while the multistep ones require two or more previous values denoted as $x_i, x_{i-1}, x_{i-2}, \ldots$, to evaluate the current iteration x_{i+1}. In both types of methods, one can also apply predictor or explicit and corrector or implicit algorithms. The explicit ones only require past iteration values to evaluate the current iteration, while the implicit algorithms require the value at the current iteration and past values. In this manner, it is common to name predictor-corrector to the implicit algorithms, and they require an explicit method to evaluate the functions at the current iteration.

Among the huge number of one-step methods, Table 2.1 lists the two basic and well-known one-step methods, namely, forward Euler, which is a predictor method, and backward Euler, which is a predictor-corrector because it requires the application of a predictor (as forward Euler) to evaluate $f(x_{i+1}, t_{i+1})$.

To reduce numerical error in the approximation of IVPs, one can apply multistep or stiff methods, as the ones described in [17], in which some of them are given in Table 2.2. In this case, the explicit or predictor method is the sixth-order Adams-Bashforth (AB6), while the implicit or predictor-corrector methods are the fourth-order Adams-Moulton (AM4) and fourth-order Gear (G4). It can be appreciated that they require more than one past value and that AM4 and G4 require the evaluation of the function at the current time step, at x_{i+1}.

Table 2.1 Numerical one-step methods

Method	Iterative equation
Forward Euler (predictor)	$x_{i+1} = x_i + hf(x_i, t_i)$
Backward Euler (predictor-corrector)	$x_{i+1} = x_i + hf(x_{i+1}, t_{i+1})$

Table 2.2 Numerical multistep methods

Method	Iterative equation
Adams-Bashforth 6 (AB6 predictor)	$x_{i+1} = x_i + \frac{h}{1440}(4277f(t_i, x_i)$ $-7923f(t_{i-1}, x_{i-1}) + 9982f(t_{i-2}, x_{i-2})$ $-7298f(t_{i-3}, x_{i-3}) + 2877f(t_{i-4}, x_{i-4})$ $-475f(t_{i-5}, x_{i-5}))$
Adams-Moulton 4 (AM4 predictor-corrector)	$x_{i+1} = x_i + \frac{h}{24}(9f(t_{i+1}, x_{i+1})$ $+19f(t_i, x_i) - 5f(t_{i-1}, x_{i-1})$ $+f(t_{i-2}, x_{i-2}))$
Gear 4 (G4 predictor-corrector)	$x_{i+1} = \frac{48}{25}x_i - \frac{36}{25}x_{i-1} + \frac{16}{25}x_{i-2}$ $-\frac{3}{25}x_{i-3} + \frac{12}{25}hf(x_{i+1})$

This book applies one-step and explicit methods to solve integer-order chaotic systems for FPGA implementation to design pseudo-random number generators (PRNGs), whose hardware descriptions are given in Chap. 3 by programming in Verilog. The randomness of the chaotic time series of the PRNGs is evaluated by the statistical methods described in Chap. 4. More elaborated methods, *ad hoc* for oscillatory systems as the chaotic ones, can also be applied as described in [18], where the authors show the FPGA-based implementation of chaotic oscillators by applying the numerical method based on trigonometric polynomials. Another issue is the adaptation of the step size of the explicit and implicit numerical methods to save time computation, which remains an open challenge.

2.2.2 Numerical Methods for Fractional-Order Chaotic Systems

As mentioned in [19], fractional-order chaotic oscillators are a hot topic nowadays, in which challenges are focused on their numerical simulation and implementation for real-world applications. For instance, different fractional-order chaotic oscillators have been implemented using different analog and digital electronic devices [11], where the field-programmable gate array (FPGA) is more popular very recently due to its capability for fast prototyping and reprogrammability. Details of the FPGA implementation of fractional-order chaotic systems are shown in [19] by solving the mathematical models applying the Grünwald-Letnikov (GL) method, which approximates the solution with reasonable exactness. The GL method is derived from fractional calculus, which was introduced since more than 300 years ago [20], but it is during this millennium that engineers, academic, and research people are introducing applications in many fields of research.

The authors in [21] provide a short description on the history of fractional calculus, which was primarily focused on the pure mathematical field, where important mathematicians like Fourier, Euler, and Laplace introduced seminal contributions. In recent years, it is quite common that an integer-order chaotic system is transformed to its fractional-order version; see, for instance, the Lorenz system in its fractional-order version in [19]. In addition, Petráš introduced the fractional-order version of several Lorenz-type systems [22], which can also be simulated by applying the GL method, and all those mathematical models have similar nonlinear functions, associated to the multiplication among state variables.

As shown in [23], fractional calculus is a generalization of the differentiation and integration to non-integer order, which is expressed by the continuous integrodifferential operator ${}_aD_t^{\alpha}$, where a and t are the limits of the operation, and α is the fractional-order value, with $\alpha \in \mathbb{R}^+$. This fractional-order operator is defined in Eq. (2.1), and by using the relation derived from the GL method, i.e., then the expression to evaluate the αth derivative when $\alpha < 1$ leads to the equation given in (2.2). In this equation h is the time step, and the binomial coefficients

$(-1)^j \binom{\alpha}{j} f(t-jh)$ can be denoted by c_j^α, for $j = 0, 1, 2 \ldots$, and can be calculated by Eq. (2.3):

$$ {}_aD_t^\alpha = \begin{cases} \frac{d^\alpha}{dt^\alpha}, & \alpha > 0, \\ 1, & \alpha = 0, \\ \int_a^t (d\tau)^\alpha, & \alpha < 0. \end{cases} \tag{2.1} $$

$$ {}_aD_t^\alpha f(t) = \lim_{h\to 0} \frac{1}{h^\alpha} \sum_{j=0}^{[\frac{t-a}{h}]} (-1)^j \binom{\alpha}{j} f(t-jh) \tag{2.2} $$

$$ c_0^\alpha = 1, \qquad c_j^\alpha = \left(1 - \frac{1+\alpha}{j}\right) c_{j-1}^\alpha \tag{2.3} $$

From the operations performed above, Eq. (2.4) describes the GL method that can be used to approximate the solution of the fractional-order derivatives (α), as shown herein in Chap. 6, where α can take the value of any positive real number in the range $0 < \alpha < 1$:

$$ g(t_k) = f(g(t_{k-1}), t_{k-1})h^\alpha - \sum_{j=v}^{k} c_j^\alpha g(t_{k-j}) \tag{2.4} $$

In Eq. (2.4), the sum is associated to the memory and hereditary properties of a fractional-order system. The sum can be truncated by applying the concept of "short-memory" principle, in which the main advantage is that it allows retaining a limited amount of information data for a relatively short period of time, so that it allows FPGA implementation with low hardware resources, as shown in this book in Chap. 6. The lower index of the sum can be defined by Eq. (2.5):

$$ v = \begin{cases} 1 & \text{when } k \le m \\ k-m & \text{when } k > m \end{cases}, \qquad m = \frac{L_m}{h}, \tag{2.5} $$

where $k = 1, 2, 3, \ldots$ and L_m is associated to the memory length. Therefore, as shown in [19], the number of binomial coefficients is equal to m and the memory beings filling places as k increases, accomplishing $k > m$. When L_m is completed, only the previous m-values of the data $x_k, x_{k-1}, \ldots, x_{k-m+1}$ are used by the summation that requires the binomial coefficients of length from c_0 to c_m. It is worth mentioning that the binomial coefficients show interesting and systematic behavior that is highlighted in Table 2.3, where it can be appreciated that their values approach to zero as m increases, so that this can be taken into account to reduce hardware resources for FPGA implementation, as shown in Chap. 6.

Table 2.3 Values of c_j^{α} as m increases for different values of the variable order α

m	1	2	3	4	5	6	7	8	9
$\alpha = 0.92$	−0.92	−0.0368	−0.0132	−0.0069	−0.0042	−0.0029	−0.0021	−0.0016	−0.0013
$\alpha = 0.93$	−0.93	−0.0325	−0.0116	−0.0060	−0.0037	−0.0025	−0.0018	−0.0014	−0.0011
$\alpha = 0.99$	−0.99	−0.0050	−0.0017	−0.00083	−0.00050	−0.00033	−0.00024	−0.00018	−0.00014

2.3 Chaotic Systems with Self-Excited and Hidden Attractors

In 1963, Lorenz introduced the first well-known chaotic system in the seminal paper titled Deterministic Nonperiodic Flow [12]. This marked the boom of chaos theory, which, as described above, becomes a branch of nonlinear system theory that has been studied intensively in recent years, not only covering the solution of integer ODEs but also the application of fractional calculus to simulate chaotic systems of non-integer or variable order. It is worth mentioning that the application of an appropriate numerical method to an integer- or fractional-order chaos generator described by nonlinear ODEs should highlight the inherent instability of the dynamical system and must lead to long-term unpredictability.

2.3.1 Chaotic Systems with Self-Excited Attractors

One of the most renowned examples of a chaotic system that generates a self-excited attractor is the well-known Lorenz system, easily identified for its intricate and butterfly-like chaotic pattern. The importance of Lorenz system is highlighted in the work titled On differences and similarities in the analysis of Lorenz, Chen, and Lu systems [24], where the authors discussed the question of the equivalence of various Lorenz-like systems and the possibility of universal consideration of reduction of such systems to the same form with the help of various transformations. This equivalence is better shown in the work [22], where the authors review not only the equivalence of integer-order chaotic systems, but also the equivalence of fractional-order Lorenz-type systems.

The Lorenz system is an autonomous system that is described by three ODEs, as shown in Eq. (2.6) [12]:

$$\begin{aligned}
\dot{x} &= \sigma(y - x),\\
\dot{y} &= \rho x - xz - y, \qquad (2.6)\\
\dot{z} &= xy - \beta z,
\end{aligned}$$

where σ, ρ, and β are constant parameters, and one can identify two nonlinear terms as the complexity introduced by two multipliers for the products xz and xy. This mathematical model can be simulated by applying one-step or multistep numerical methods [15]. However, a challenge is the selection of an adequate method to approximate the solution of a given problem described by nonlinear ODEs. In addition, the estimation of the time step is a challenge to guarantee the lowest error with respect to an exact solution and to increase the operating frequency of an application based on chaotic systems, as detailed in [11, 25]. In fact, choosing an inappropriate numerical method and/or step size, the approached solution may diverge, be erroneous, or give rise to undesired computational effects like superstability. For instance, as shown in [11], the simplest method to solve ODEs is the well-known forward Euler. However, this is a one-step method, which provides higher error compared to the application of a multistep, stiff, or special method. To guarantee low error for the simulation of nonlinear ODEs, the majority of people applies the fourth-order Runge-Kutta method, which allows a higher time step compared to forward Euler. However, as shown in [25], without supporting analysis, yet a numerical simulation of a simple dynamical system may engender the appearance of spurious solutions.

Forward Euler and Runge-Kutta methods are good options to simulate dynamical systems like the one given in Eq. (2.6), but if one must ensure better exactness, one can apply numerical methods *ad hoc* to problems having oscillatory characteristics. For example, the authors in [18] implemented chaotic systems into an FPGA by applying a method based on trigonometric polynomials, which provide better accuracy than forward Euler and Runge-Kutta. This method can also be applied to the 3D Lorenz-type systems described in [22]. In fact, the authors in [26] show the experimental observation of seven 3D Lorenz-type systems that are simulated using the Verilog hardware description language and implemented in an FPGA from the application of the forward Euler method.

Other well-known 3D chaotic systems with self-excited attractors are the Chen and Rössler systems that are modeled by the ODEs given in Eqs. (2.7) and (2.8), respectively:

$$\begin{aligned} \dot{x} &= a(y - x), \\ \dot{y} &= (c - a)x - xz + cy, \\ \dot{z} &= xy - bz. \end{aligned} \tag{2.7}$$

$$\begin{aligned} \dot{x} &= -(y + z), \\ \dot{y} &= x + ay, \\ \dot{z} &= z(x - c) + b. \end{aligned} \tag{2.8}$$

In Eqs. (2.7) and (2.8), a, b, c are constant parameters. It is important to mention that choosing the appropriate parameter values in Eqs. (2.6), (2.7), and (2.8) can lead to generate chaotic time series, in which dynamics can be measured by evaluating their associated Lyapunov exponents. For example, Table 2.4 shows the parameter values and initial conditions of these 3D systems that can generate self-excited attractors.

Table 2.4 Parameter values and initial conditions to simulate 3D chaotic systems

Chaotic system	Parameter values	Initial conditions
Lorenz (2.6)	$\sigma = 10, \rho = 28, \beta = 8/3$	$(x(0), y(0), z(0)) = (10, 10, 10)$
Chen (2.7)	$a = 35, b = 3, c = 28$	$(x(0), y(0), z(0)) = (0.5, 1, 5)$
Rössler (2.8)	$a = 0.2, b = 0.2, c = 5.7$	$(x(0), y(0), z(0)) = (40, -20, 30)$

Using the values listed in Table 2.4, one can analyze the 3D chaotic systems to find their equilibrium points and to evaluate their Lyapunov exponents. For instance, Table 2.5 shows the equilibrium points, eigenvalues, and Lyapunov exponents for the Lorenz and Chen systems.

The Kaplan-Yorke dimension D_{KY} is a quite useful characteristic for quantifying chaotic dynamics and it can be evaluated from the Lyapunov exponents, so that for an n-dimensional system it is evaluated by Eq. (2.9), where $LE_1, \ldots, LE_n$ are Lyapunov exponent values ordered from the highest to the lowest one:

$$D_{KY} = (n-1) + \frac{LE_1 + LE_2 + LE_{n-1}}{|LE_n|} \tag{2.9}$$

2.3.2 Chaotic Systems with Multistability and Hidden Attractors

Extensive research has shown that multistability and hidden attractors can appear in different nonlinear systems. In this subsection, 3D, 4D, and 5D chaotic systems showing multistability and hidden attractors are described.

A 3D memristive chaotic system already implemented in FPGA and showing multistability and initial-offset boosting dynamics has been introduced in [27], wherein the mathematical model is given in Eq. (2.10):

$$\begin{cases} \dot{x} = ayz, \\ \dot{y} = x - y, \\ \dot{z} = k - b(\gamma xy - \delta xy^3). \end{cases} \tag{2.10}$$

This model embeds a memristor function and generates chaotic behavior by setting $a = 3.5$, $k = 1.2$, $b = 1.8$, $\gamma = 2$, and $\delta = 0.1$. Further, its stability analysis is based on the evaluation of the equilibrium points and their corresponding eigenvalues that are listed in Table 2.6. The eigenvalues of the system are determined through the Jacobian matrix given in Eq. (2.11):

$$J(x^\star, y^\star, z^\star) = \begin{pmatrix} 0 & az^\star & ay^\star \\ 1 & -1 & 0 \\ -b(\gamma y^\star - \delta y^{\star 3}) & -b(\gamma x^\star - 3\delta x^\star y^{\star 2}) & 0 \end{pmatrix}. \tag{2.11}$$

As shown in [27], the equilibrium points EP_1 and EP_3 are unstable saddle points of index one, with λ_1 as a positive real number and $\lambda_2 = 0$ and λ_3 a negative real number. The other equilibrium points EP_2 and EP_4 are stable points because they have $\lambda_1 = 0$ and two complex conjugate eigenvalues with negative real part.

This 3D memristive chaotic system can be simulated by applying numerical methods. For example, by applying forward Euler the discretized equations are

Table 2.5 Equilibrium points, eigenvalues, and Lyapunov exponents of Lorenz and Chen systems

			Lyapunov exponents		
Chaotic system	Equilibrium points	Eigenvalues	L_1	L_2	L_3
Lorenz	$(0, 0, 0)$	$\{-22.828, 11.828, -2.667\}$	− 22.828	11.828	− 2.667
	$(-6\sqrt{2}, -6\sqrt{2}, 27)$	$\{-13.855, 0.094 + 10.194i, 0.094 - 10.194i\}$	− 13.855	0.094	0.094
	$(6\sqrt{2}, 6\sqrt{2}, 27)$	$\{-13.855, 0.094 + 10.194i, 0.094 - 10.194i\}$	− 13.855	0.094	0.094
Chen	$(0, 0, 0)$	$\{-30.836, 23.836, -3.000\}$	− 30.836	23.856	− 3.000
	$(-3\sqrt{7}, -3\sqrt{7}, 21)$	$\{-18.428, 4.214 + 14.885i, 4.214 - 14.885i\}$	− 18.428	4.214	4.214
	$(3\sqrt{7}, 3\sqrt{7}, 21)$	$\{-18.428, 4.214 + 14.885i, 4.214 - 14.885i\}$	− 18.428	4.214	4.214

Table 2.6 Equilibrium points and corresponding eigenvalues from Eq. (2.10)

Equilibrium points	Eigenvalues		
$EP_1 = (0.58, 0.58, 0)$	$\lambda_1 = 1.01$;	$\lambda_2 = 0$;	$\lambda_3 = -2.01$
$EP_2 = (-0.58, -0.58, 0)$	$\lambda_1 = 0$;	$\lambda_2 = -0.5 + 1.34i$;	$\lambda_3 = -0.5 - 1.34i$
$EP_3 = (4.43, 4.43, 0)$	$\lambda_1 = 3.47$;	$\lambda_2 = 0$;	$\lambda_3 = -4.47$
$EP_4 = (-4.43, -4.43, 0)$	$\lambda_1 = 0$;	$\lambda_2 = -0.5 + 3.90i$;	$\lambda_3 = -0.5 - 3.90i$

shown in Eq. (2.12), which can be synthesized into an FPGA, as shown in Chap. 5:

$$\begin{aligned} x[n+1] &= x[n] + h(ay[n]z[n]), \\ y[n+1] &= y[n] + h(x[n] - y[n]), \\ z[n+1] &= z[n] + h(k - b\gamma x[n]y[n] - b\delta x[n]y[n]^3), \end{aligned} \tag{2.12}$$

where, as mentioned above, the constant parameters can be set to $b = 1.8$, $\gamma = 2$, $\delta = 0.1$, and the parameters a and k can be varied to form different attractors, as shown in [27].

A 4D hyperjerk system with a half line equilibrium, showing multistability, period doubling reversals, and antimonotonocity, has been introduced in [28], which has also been implemented in FPGA an applied to image encryption. The mathematical model of this 4D chaotic system is given in Eq. (2.13):

$$\begin{cases} \dot{x} = y, \\ \dot{y} = z, \\ \dot{z} = w, \\ \dot{w} = -x - |x| - ay - bw - xz. \end{cases} \tag{2.13}$$

Bao et al. showed that this hyperchaotic system generates chaotic behavior by setting $a = 4$ and $b = 2$ with the initial conditions $(x_0, y_0, z_0, w_0) = (0.02, 0.02, 0.02, 0.02)$ [28]. In Eq. (2.13), it is straightforward to see that the system has the equilibrium set $S = \{(x, y, z, w) | x \leq 0, y = 0, z = 0, w = 0\}$, which is a half line consisting of the nonpositive x-axis in $\mathbf{R}^4$. This 4D hyperchaotic system is described by Verilog and synthesized into an FPGA to design a PRNG, as shown in Chap. 5.

Another case study that is considered in this book to design a PRNG is a hyperchaotic system presenting extreme multistability and extreme events with hidden attractors. This is a 5D chaotic system [29, 30] consisting of six parameters and five state variables. The mathematical model introduced by the authors in [30] is given in Eq. (2.14):

$$\begin{cases} \dot{x} = ax - yz - u, \\ \dot{y} = -by + xz - zw, \\ \dot{z} = -cz + xy, \\ \dot{w} = d(y - w), \\ \dot{u} = ex + s, \end{cases} \tag{2.14}$$

where the system generates hyperchaotic behavior by setting: $a = 3.0, b = 2.5, c = 1.0, d = 4.0, e = 3.5, s = 0.1$ and the step size to $h = 0.001$. This 5D hyperchaotic system can be simulated by setting the initial conditions to $(x_0, y_0, z_0, w_0, u_0) = (1.0, 1.0, 1.0, 1.0, 1.0)$.

The authors in [30] performed the FPGA implementation of the autonomous five-dimensional (5D) system with offset boosting. The article shows that the 5D hyperchaotic system displays several interesting features and behaviors, including Hopf bifurcation; reverse period doubling route to chaos; intermittency route to chaos; bistability phenomenon; periodic, quasi-periodic, and one- and double-scroll chaotic attractors; and partial amplitude control of its signals. The experimental results were applied to a sound encryption scheme using a chaos masking technique. This means that the generated binary sequences must be random, as they can be evaluated by performing TestU01 and NIST statistical tests, as shown in Chap. 4. This 5D hyperchaotic system is described under Verilog and synthesized into an FPGA to design a PRNG, as shown in Chap. 5.

All these 3D, 4D, and 5D systems with self-excited and hidden attractors have been implemented in embedded hardware, in their respective references mentioned above, and are case studies in this book to implement PRNGs, showing their mathematical analysis, Verilog description, and FPGA implementation along cryptographic applications.

2.4 Fractional-Order Chaotic Systems

The fractional-order (FO) Lorenz's chaotic oscillator and other 3D FO Lorenz-type systems are analyzed in [22]. The FO Lorenz system is described herein in Eq. (2.15), and it can generate chaotic behavior by using the initial conditions $(x(0), y(0), z(0) = (0.1, 0.1, 0.1)$, using the fractional orders $(\alpha_1, \alpha_2, \alpha_3) = (0.995, 0.995, 0.995)$, and by setting the parameter values to $\sigma = 10$, $\rho = 28$ and $\beta = \frac{8}{3}$. As one can infer, choosing the parameter values of a FO chaotic system can be quite difficult in a trial and error fashion. Fortunately, the authors in [31] showed the application of metaheuristics to perform the optimization of integer and FO chaotic systems, so that practically all integer-order chaotic and hyperchaotic systems can be transformed to be of FO type, and the optimization algorithms can find the appropriate parameter values to generate and improve chaotic behavior:

$$
\begin{aligned}
D_*^{\alpha 1} x &= \sigma(y - x), \\
D_*^{\alpha 2} y &= x(\rho - z - y), \\
D_*^{\alpha 3} z &= xy - \beta z.
\end{aligned} \tag{2.15}
$$

Other types of FO chaotic systems are the ones based on memristor, as the one introduced in [27] that shows multistability and initial-offset boosting dynamics, in which a mathematical model is given in Eq. (2.10). This type of integer-order systems can be modeled as a FO one and the solution can be approximated by applying methods as the ones described in Sect.2.2.2. In other cases of memristive systems, one can add time delay. For example, let us consider the memristor model shown in Eq. (2.16) [32]:

$$
\begin{cases}
i_m = W(\varphi) v_m, \\
W(\varphi) = \gamma - \delta\varphi^2, \\
\dot{\varphi} = v_m - \varphi,
\end{cases} \tag{2.16}
$$

where $\gamma = 1$ and $\delta = 0.1$. This model can be expressed as a first-order nonlinear delay equation, with a delay constant $\tau \in R^+$, as given in Eq. (2.17):

$$
\dot{x} = -ax(t) + bW(x_\tau), \tag{2.17}
$$

where $W(x_\tau) = \gamma - \delta x_\tau^2$, and thus, the integer-order version of Eq. (2.17) can be rewritten by Eq. (2.18):

$$
\dot{x} = -ax(t) + b(\gamma - \delta x_\tau^2), \tag{2.18}
$$

with $x_\tau = x(t - \tau)$. The FO version can be described by Eq. (2.19):

$$
D_*^{\alpha} x = -ax(t) + b(\gamma - \delta x_\tau^2), \tag{2.19}
$$

which can be solved by applying a numerical method for FO systems including time delay and the short memory principle, as shown below:

The numerical method introduced in [33] is described in Eq. (2.20), which can be applied to solve the fractional-order memristive system of delay differential equations:

$$
\omega_0 \chi_n + \sum_{j=1}^{k-1} \omega_j \chi_{k-j} = \omega_n \chi_0 + h^{\mu} \Gamma(2 - \mu) \psi(t_n, \chi_n, \chi_{\tau n}), \tag{2.20}
$$

where μ denotes the fractional order of the dynamical system, n is the nth iteration, and k is the number of steps. The coefficients ω_0, ω_j, ω_n are set to the values given in Eq. (2.21):

$$\begin{cases} \omega_0 = \gamma_0 = 1, \\ \omega_j = \gamma_j - \gamma_{j-1}, \ j = 1, 2, ..., n-1, \\ \omega_n = \gamma_{n-1} \end{cases} \tag{2.21}$$

where $\gamma_j = (j+1)^{1-\mu} - j^{1-\mu}$.

Equation (2.20) describes a numerical scheme involving a summation of n elements, where n represents the value of the iteration being calculated at that moment. Thus, it is necessary to have a history of all the approximations χ that have already been calculated. However, thinking on hardware implementation, the number of elements being added must be shortened, e.g., by applying the short-term memory principle described in Sect.2.2.2. In this manner, considering the behavior of $f(t)$ only in the values close to the nth iteration, i.e., in the interval $[t - L_m, t]$, then the memory length L_m can be associated to Eq. (2.22) [34]:

$$_aD_t^{\mu} f(t) \approx_{t-L_m} D_t^{\mu} f(t), \ (t > a + L_m), \tag{2.22}$$

so that the number of elements in the summation in Eq. (2.20) is always no greater than $[L_m/h]$. Therefore, if $f(t) \leq M$ for $a \leq t \leq b$, the necessary memory length L_m to achieve the required precision ϵ can be determined through the inequality given in Eq. (2.23):

$$L_m \geq \left(\frac{M}{\epsilon |\Gamma(1-\mu)|} \right). \tag{2.23}$$

Given the above, the adaptation of the numerical scheme described in Eq. (2.20) becomes

$$\omega_0 \chi_n + \sum_{j=1}^{m-1} \omega_j \chi_{m-j} = \omega_m \chi_0 + h^{\mu} \Gamma(2-\mu) \psi(t_n, \chi_n, \chi_{\tau n}), \tag{2.24}$$

with $m = L_m/h$, and the coefficients of Eq. (2.21) only take values in the range $j = 1, 2, \ldots, m-1$.

The simulation of the FO memristor model with time delay can be observed in Fig. 2.1, where the temporal evolution does not reach a value higher than approximately 4, so that $M = 4.5$ can be established to use it in Eq. (2.23).

Figure 2.2 shows the phase diagrams using the short-term memory principle for $L_m = 1.28$, which results in an accuracy of $\epsilon_{0.96} = 0.145$ and $\epsilon_{0.98} = 0.071$ for $\mu = 0.98$ and $\mu = 0.96$, respectively.

As $\omega_0 = 1$ in Eq. (2.21), then the numerical method given in Eq. (2.24) can be rewritten as described in Eq. (2.25), which is suitable to be implemented through digital hardware, as shown in this book in Chap. 6, where the FPGA synthesis is

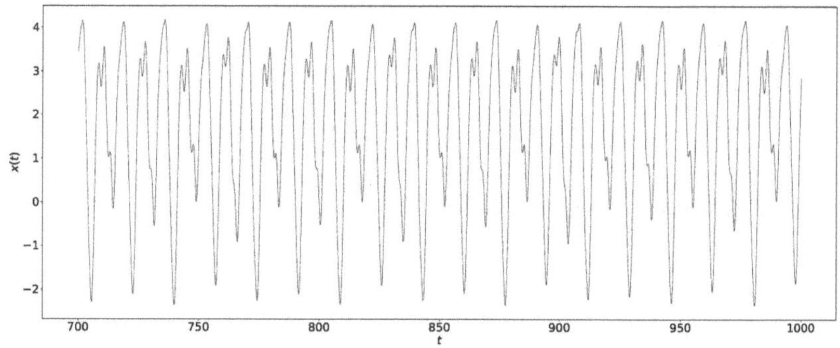

Fig. 2.1 Temporal evolution of the memristor model

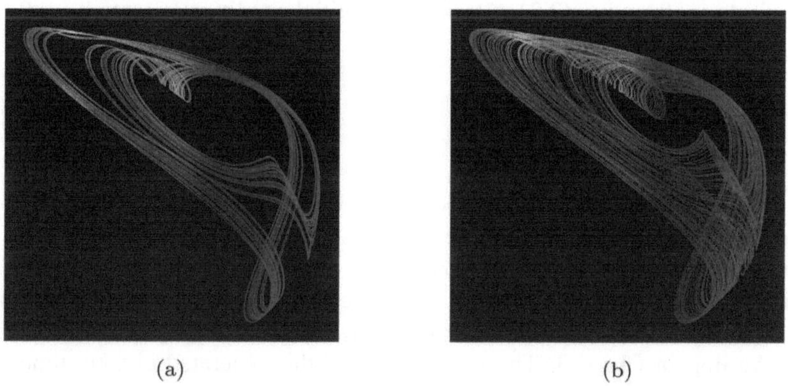

Fig. 2.2 Portraits of the time delay system for $\tau = 3.5$; $a = 1.0$; $b = 4.5$; $h = 0.01$; $x_0 = 0.0$, varying the fractional order to (**a**) $\mu = 0.96$ and (**b**) $\mu = 0.98$

used to design a PRNG:

$$\begin{aligned} x_n &= \omega_m x_0 + h^{\mu}\Gamma(2-\mu)\psi(t_{n-1}, x_{n-1}, x_{\tau n-1}) - \sum_{j=1}^{m-1} \omega_j x_{m-j}, \\ x_{\tau n} &= x_{n-n_\tau}. \end{aligned} \tag{2.25}$$

The function ψ and the variable $x_{\tau n}$ have different behaviors depending on n and n_τ values. ψ is defined by Eq. (2.26):

$$\psi(t_n, x_n, x_{\tau n}) = -a x_n + b(F_{x\tau n}), \tag{2.26}$$

where n_τ is evaluated by using the step size h and delay time τ, as $n_\tau = \frac{\tau}{h}$, and $F_{x\tau n}$ is given in Eq. (2.27):

$$F_{x\tau n} = \begin{cases} 0.55 & \text{if } n \le n_\tau, \\ \gamma - \delta x_{\tau n}^2 & \text{if } n > n_\tau, \end{cases} \tag{2.27}$$

and $x_{\tau n}$ can take the values given in Eq. (2.28):

$$x_{\tau n} = \begin{cases} 0 & \text{if } n \le n_\tau, \\ x_{n-n_\tau} & \text{if } n > n_\tau. \end{cases} \tag{2.28}$$

This numerical method solves the memristive-based chaotic system with delay time by setting $h = 0.01$, $\tau = 3.5$, $L_m = 1.28$, $m = 128$, $n_\tau = 350$, and taking 128 coefficients ω_j from Eq. (2.21). Besides, the initial condition is set to $x_0 = 0$.

2.5 Conclusion

This chapter showed the mathematical models of several integer- and fractional-order chaotic systems. Some one-step and multistep numerical methods have been described to solve chaotic systems consisting of ODEs. In the case of a fractional-order system, a numerical method was described to simulate a 3D memristive fractional-order chaotic system with time delay. These chaotic systems are described under Verilog in Chap. 3. The randomness of the generated chaotic time series can be evaluated by applying TestU01 and NIST statistical tests, as it will be shown in Chap. 4. Chapters 5 and 6 show the FPGA synthesis of PRNGs based on chaotic maps and integer- and fractional-order chaotic systems. Finally, from the mathematical model of a chaotic system, one can design a PRNG whose application is useful in the authentication of encrypted color images, as it will be shown in Chap. 7.

References

1. S.H. Strogatz, *Nonlinear Dynamics and Chaos with Student Solutions Manual: With Applications to Physics, Biology, Chemistry, and Engineering* (CRC Press, Boca Raton, 2018)
2. J.C. Sprott, *Chaos and Time-Series Analysis* (Oxford University Press, Oxford, 2003)
3. T. Yang, L.O. Chua, Impulsive stabilization for control and synchronization of chaotic systems: theory and application to secure communication. IEEE Trans. Circuits Syst. I Fund. Theory Appl. **44**(10), 976–988 (1997)
4. D. Clemente-Lopez, J. de Jesus Rangel-Magdaleno, J.M. Muñoz-Pacheco, A lightweight chaos-based encryption scheme for iot healthcare systems. Internet Things **25**, 101032 (2024)

5. Y. Liang, G. Wang, G. Chen, Y. Dong, D. Yu, H. Ho-Ching Iu, S-type locally active memristor-based periodic and chaotic oscillators. IEEE Trans. Circuits Syst. I Regular Papers **67**(12), 5139–5152 (2020)
6. K. Yang, J. Joshua Yang, R. Huang, Y. Yang, Nonlinearity in memristors for neuromorphic dynamic systems. Small Sci. **2**(1), (2022). https://doi.org/10.1002/smsc.202100049
7. C.K. Volos, I.M. Kyprianidis, I.N. Stouboulos, Experimental investigation on coverage performance of a chaotic autonomous mobile robot. Robot. Auton. Syst. **61**(12), 1314–1322 (2013)
8. H.E. Kiran, A. Akgul, O. Yildiz, E. Deniz, Lightweight encryption mechanism with discrete-time chaotic maps for internet of robotic things. Integration **93**, 102047 (2023)
9. F.A. Hossam, E. Mohamed, W. El-Shafai, Cancelable biometric authentication system based on hyperchaotic technique and fibonacci q-matrix. Multimedia Tools Appl., 1–39 (2024)
10. N. Charalampidis, A. Iatropoulos, C. Volos, Chaos based speech encryption using microcontroller. Integration **95**, 102128 (2024)
11. E. Tlelo-Cuautle, A.D. Pano-Azucena, O. Guillén-Fernández, A. Silva-Juárez, *Analog/Digital Implementation of Fractional Order Chaotic Circuits and Applications* (Springer, New York, 2020)
12. E.N. Lorenz, Deterministic nonperiodic flow. J. Atmos. Sci. **20**(2), 130–141 (1963)
13. G. Chen, T. Ueta, Yet another chaotic attractor. Int. J. Bifur. Chaos **9**(7), 1465–1466 (1999)
14. O.E. Rossler, An equation for continuous chaos. Phys. Lett. A **57**(5), 397–398 (1976)
15. J.D. Lambert, Computational Methods in Ordinary Differential Equations (Wiley, Hoboken, Nueva Jersey, EUA, 1973)
16. A. Wolf, J.B. Swift, H.L. Swinney, J.A. Vastano, Determining lyapunov exponents from a time series. Phys. D Nonlinear Phenomena **16**(3), 285–317 (1985)
17. L.O. Chua, Computer-Aided Analysis of Electronic Circuits. Prentice-Hall series in electrical and computer engineering (1975)
18. A.D. Pano-Azucena, E. Tlelo-Cuautle, G. Rodriguez-Gomez, L.G. de la Fraga, FPGA-based implementation of chaotic oscillators by applying the numerical method based on trigonometric polynomials. AIP Adv. **8**(7), 75217 (2018)
19. A.D. Pano-Azucena, B. Ovilla-Martinez, E. Tlelo-Cuautle, J.M. Muñoz-Pacheco, L.G. de la Fraga, Fpga-based implementation of different families of fractional-order chaotic oscillators applying grünwald–letnikov method. Commun. Nonlinear Sci. Numer. Simul. **72**, 516–527 (2019)
20. H.G. Sun, Y. Zhang, D. Baleanu, W. Chen, Y.Q. Chen, A new collection of real world applications of fractional calculus in science and engineering. Commun. Nonlinear Sci. Numer. Simul. **64**, 213–231 (2018)
21. H. Zhu, C. Zeng, A novel chaotification scheme for fractional system and its application. J. Comput. Appl. Math. **339**, 275–284 (2018)
22. I. Petráš, The fractional-order lorenz-type systems: A review. Fract. Calculus Appl. Anal. **25**(2), 362–377 (2022)
23. O.I. Tacha, J.M. Munoz-Pacheco, E. Zambrano-Serrano, I.N. Stouboulos, V.-T. Pham, Determining the chaotic behavior in a fractional-order finance system with negative parameters. Nonlinear Dynam. **94**(2), 1303–1317 (2018)
24. G.A. Leonov, N.V. Kuznetsov, On differences and similarities in the analysis of lorenz, chen, and lu systems. Appl. Math. Comput. **256**, 334–343 (2015)
25. J.D. Skufca, Analysis still matters: A surprising instance of failure of runge–kutta–felberg ode solvers. SIAM Rev. **46**(4), 729–737 (2004)
26. J.D. Tavizon-Aldama, J.D. Rodriguez-Munoz, E. Tlelo-Cuautle. Experimental observation of chaotic attractors from the fpga implementation of 3d chaotic systems. IETE J. Educ. **65**(1), 6–14 (2024)
27. Z. Kotadai, C. Fischer, J.D. Rodríguez-Muñoz, E. Tlelo-Cuautle, E.J. Yves, Multistability and initial-offset boosting dynamics in a new 3d memristive chaotic system with fpga implementation. Eur. Phys. J. Plus **139**(1), 70 (2024)
28. A. Sambas, M. Miroslav, S. Vaidyanathan, B. Ovilla-Martínez, E. Tlelo-Cuautle, A.A. Abd El-Latif, B. Abd-El-Atty, K. Benkouide, T. Bonny, A new hyperjerk system with a half line

equilibrium: Multistability, period doubling reversals, antimonotonocity, electronic circuit, fpga design, and an application to image encryption. IEEE Access **12**, 9177–9194 (2024)
29. A. Ahmadi, S. Parthasarathy, N. Pal, K. Rajagopal, S. Jafari, E. Tlelo-Cuautle, Extreme multistability and extreme events in a novel chaotic circuit with hidden attractors. Int. J. Bifurcat. Chaos **33**(07), 2330016 (2023)
30. S. Takougang Kingni, K. Rajagopal, S. Çiçek, A. Srinivasan, A. Karthikeyan, Dynamic analysis, fpga implementation, and cryptographic application of an autonomous 5d chaotic system with offset boosting. Front. Inf. Technol. Electronic Eng. **21**, 950–961 (2020)
31. E. Tlelo-Cuautle, L.G. De La Fraga, O. Guillén-Fernández, A. Silva-Juárez, *Optimization of Integer/Fractional Order Chaotic Systems by Metaheuristics and Their Electronic Realization* (CRC Press, Boca Raton, 2021)
32. Z. Kotadai, E.J. Yves, C. Fischer, J.D. Rodríguez-Muñoz, M.F. Moreno-Lopez, E. Tlelo-Cuautle, J. De Dieu Nkapkop, Fractional order 1d memristive time-delay chaotic system with application to image encryption and fpga implementation. Math. Comput. Simul. (2024). https://doi.org/10.1016/j.matcom.2024.07.035
33. M. Kumar, An Efficient Numerical Scheme for Solving a Fractional-Order System of Delay Differential Equations. Int. J. Appl. Comput. Math. **8**(5), 1–18 (2022)
34. I. Podlubny, *Fractional Differential Equations*, vol. 148 (Academic Press, Cambridge, 1998)

Chapter 3
Verilog Descriptions of Digital Blocks to Synthesize Chaotic Maps and Systems

3.1 Introduction

Up to now, chaotic systems have demonstrated their usefulness in the development of hardware security applications, as the one described in this book for the design of PRNGs. As their main characteristic is the high sensitivity to initial conditions [1], then chaotic systems in their integer or fractional-order versions find applications in cryptography and authentication of data, such as images, as is shown in this book in Chap. 7.

Integer-order chaotic systems are modeled by ordinary differential equations (ODEs), and fractional-order chaotic systems are modeled by fractional-order differential equations (FODEs). ODEs and FODEs can be solved by applying numerical methods, which leads to discretized equations, as is shown in Chap. 2. If one applies explicit methods, as done in this book, the discretized equations evaluate state variables at iteration $n+1$, as a function of one previous iteration n by applying one-step methods, or as functions of more than one previous iterations $n, n-1, n-2, \ldots$, by applying multistep methods. This last case also applies for fractional-order chaotic systems by approximating the fractional-order derivative by summation of terms at previous iterations, as is shown in Sect. 2.4, in Chap. 2, for the fractional-order memristor system with time delay. The resulting discretized equations can then be implemented with digital devices as the field-programmable gate array (FPGA), which is suitable to synthesize equations consisting of adders, subtractors, multipliers, multiplexers, registers, and so on. These blocks require to estimate the number of bits that are necessary to process the amplitudes of the signals for each state variable. However, one must estimate the maximum amplitude that may arise when multiplying two state variables, as is detailed in Chap. 5. For example, if one chaotic system has amplitude in the state variables $x = 50$ and $y = 100$, then there exists the possibility that in one discrete time the multiplication can generate a maximum amplitude of $xy = 5000$, so that this value must be represented in binary. As is shown in Chap. 4, using computer arithmetic in fixed-point format is quite

L. G. de la Fraga et al., *Random Number Generators*,
https://doi.org/10.1007/978-3-031-82865-2_3

suitable for FPGA implementation. In this manner, since the goal of this book is the FPGA implementation, the number 5000 can be described in fixed-pint format by using 13 bits ($2^{13} = 8192$), because by using 12 bits the value is lower than 5000 ($2^{12} = 4096$). If one includes one bit to represent the sign and one decides to use 32 bits to design each block (adder, subtractor, and so on), then the 32 bits can be distributed in fixed-point format as 1 bit for the sign, 13 bits for the integer part, and 18 bits for the decimal part. This computer arithmetic is detailed in Chaps. 5 and 6, for the FPGA implementation of chaotic maps and integer and fractional-order chaotic systems.

The synthesis of a chaotic system into an FPGA requires to describe the blocks by using a hardware description language (HDL), which depends on the type of FPGA. In this book, the experiments are developed using an FPGA from ALTERA-Intel, who requires to describe the blocks by using Verilog with the compiler Quartus II, available for windows at https://www.intel.com/content/www/us/en/software-kit/666221/intel-quartus-ii-web-edition-design-software-version-13-1-for-windows.html?wapkw=Quartus%20II, and Linux at https://www.intel.com/content/www/us/en/software-kit/666220/intel-quartus-ii-web-edition-design-software-version-13-1-for-linux.html?wapkw=Quartus%20II.

As the FPGA has the advantage for fast prototyping, one can test the suitability of chaotic models to generate random sequences. This process is performed in this book beginning with the Verilog descriptions of chaotic systems, following with the FPGA synthesis. A key point is the use of the chaotic systems to implement a PRNG, who is validated by performing NIST and TestU01 statistical tests, as is shown in Chap. 4, and finally, the designed PRNG is applied to encrypt and authenticate RGB images, as is shown in Chap. 7. The process of simulating the chaotic systems under Verilog for the compiler Quartus II requires that one think in parallel to save hardware resources during the FPGA synthesis [2]. The basic blocks for the implementation of operations like addition, subtraction, multiplication, and sequential operations are given in this chapter, whose Verilog descriptions are detailed.

As mentioned above, the first step to design digital blocks is the evaluation of the amplitudes of the state variables and their operations that may generate large numbers, which can require more bits for their digital representation. Another issue is the computer arithmetic, as the fixed-point notation that is applied in this book. This notation uses one bit to represent the sign of a number, and the rest of the bit are divided to represent the integer numbers and decimal part. As the majority of chaotic systems available in current literature have been implemented on FPGA using 32 or 64 bits, the following blocks are described under Verilog by using a fixed-point format. The process of assigning the number of bits for the implementation of a chaotic system can be revised in works like the one in [3].

3.2 Verilog Descriptions of Arithmetic Blocks

Verilog is a hardware description language that allows the description of digital systems in a hierarchical fashion, where the architecture of the basic block is described by using the reserved word called *module*. Each module can be seen as a black box or block, which can have inputs, outputs, and pins for connecting a clock (CLK) or pins for connecting controlling signals. The block can perform logic or arithmetic operations in a combinational or sequential fashion, which is associated to the behavior [4]. The programming of a block is then described between the reserved words **module** and **endmodule**.

The organization of the structure of the Verilog codes is performed by instantiation of modules, which consists of creating copies of a module within another of higher hierarchy. This instantiation process allows the creation of complex blocks from the description and interconnection of basic blocks. As one can find different ways of instantiation processes, this book uses Verilog descriptions in which bigger and complex designs are built by integrating multiple modules in a hierarchical manner. Basically, basic modules are instantiated within modules and ports and the instances can be connected with other signals inside the parent module. For instance, the connection method called ordered port can be revised in [5].

Verilog includes reserved words that define the ports and the architecture and behavior of a module. For example, the words **input** and **output** are used to establish the buses that describe the input and output of data of a block. The behavior of a block is established by using the reserved words **assign**, **initial**, and **always**. The continuous assignation uses the data type with the reserved word **wire** to specify combinational logic. The behavior of a process can apply **initial** and **always** to describe combinational or sequential blocks through the use of reserved words, such as **reg**, which is in charge of the data storage. The reserved word **signed** can be used to describe blocks that work with positive and negative numbers by using computer arithmetic of 2's complement and, in addition, enables the extension of the most significant bit in operations like multiplication and shift. Another reserved word is **parameter**, which allows parameterizing blocks to modify their instantiation properties. These properties include the size of the data buses, the initialization of values of some parameters, among others.

This book uses Verilog to describe chaotic systems and their applications to design a PRNG for encryption and authentication of RGB images. As is shown in Chap. 2, chaotic systems are modeled by ODEs and FODEs whose discretization by applying numerical methods generate iterative equations suitable for FPGA implementation. In the majority of cases, the equations of a chaotic system can be described by blocks in charge of performing addition, subtraction, and multiplication operations. These blocks are combinational and do not require a CLK signal to produce a result at their output. However, thinking on FPGA implementations, these blocks require a propagation time to produce a result, which depends on the fabrication technology of the device. In this manner, the series connection of $nblock$ combinational blocks increases the time to provide an output, which is equal to

$nblock \times$ propagation time. This produces a reduction of the frequency of operation that controls sequential blocks, as is shown in [6]. One solution to increase the frequency of operation consists on converting the arithmetic blocks to include CLK pin, thus behaving as a sequential block. This is performed by adding a register with parallel input and output at the output of each combinational block.

The following subsections show the Verilog description of the combinational and sequential blocks that will be required to describe a chaotic system. Each block is sketched by a diagram that shows the pines and CLK. The first block to be described is the parallel input and parallel output, which is used in the combinational blocks performing addition, subtraction, and multiplication, in order to increase the operation frequency of a chaotic system, as described in Chaps. 5 and 6.

3.2.1 Parallel-Input, Parallel-Output (PIPO) Register

A parallel-input, parallel-output (PIPO) register is a sequential block that is designed using type-D flip-flops, and its purpose in this book is the storage of data, as temporal ones. The input and output buses have the same size (n bits) and include three control pines: One pin is assigned to drive the clock (CLK), one pin to assign a reset (RST), and one more to enable (ENB) the behavior. This module performs the following operation: If the ENB pin is in a high state (logic 1 value) and the RST pin in a low one (logic 0 value), then each rising (or falling) edge in the CLK signal will transfer the data from the input bus to the output bus. Otherwise, if ENB pin is in a low state (logic 0 value) and the RST pin in a low one (logic 0 value), the output bus of the PIPO register will hold the present data at its output, independently of the data available at the input bus and of the transitions of the clock. Besides, if pin RST is in a high state (logic 1 value), the output bus of the PIPO register will load the initial data values without the necessity of waiting for a rising (or falling) edge in the clock signal. Figure 3.1 shows the synthesis of the PIPO register, detailing the digital hardware that consists of n-bit input (I) and output (O) busses, type-D flip-flops, and three control pines associated to the signals called clock (CLK), reset (RST), and enable (ENB).

The block description of the PIPO register is shown in Fig. 3.1, and it can be described under Verilog, as detailed in Listing 3.1. This code is in parameterized format to allow the modification of the number of bits n for the input and output buses and the initial values ($init$) when the RST pin is activated. If during an instantiation process of this PIPO register, the values of the parameters n and $init$ are not provided or established, then the register will use predetermined values, which are set to $n = 32$ bits and init $=0$.

The Verilog description sketched in Listing 3.1 imposes that the use of the PIPO register must have n-bit length to be connected to other blocks. As mentioned above, the design of combinational blocks like the ones required to synthesize chaotic systems, e.g., adders, subtractors, and multipliers, can be converted to sequential ones just by connecting a PIPO register at their output buses, which can help to

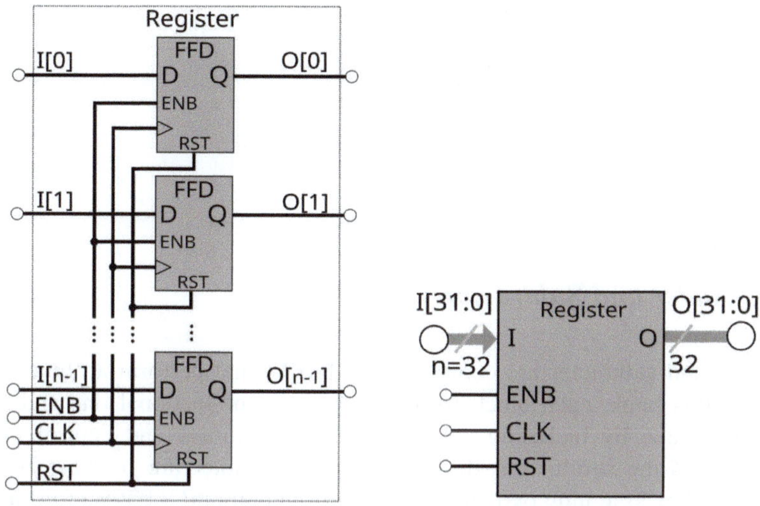

(a) (b)

Fig. 3.1 Parallel-input, parallel-output (PIPO) register: (**a**) block description of the digital hardware consisting of n-bit input-output busses and three control pines, namely, clock (CLK), reset (RST), and enable (ENB). (**b**) Abstract block description of PIPO register showing input and output buses of 32 bits, CLK, RST, and ENB pines

increase the operating frequency. In this case, each block will require to control the CLK pin, so that the design of finite-state machines (FSMs) must consider such process to estimate the appropriate number of clock cycles (CCs) that are required to evaluate a mathematical expression, as is shown in the following chapters.

Listing 3.1 Verilog description of PIPO register

```
/*-----------------------------------------------------------------
---Project: parallel-in/parallel-out register of n bits        ---
---Inputs: RST - sets the values of the output bus with the
    initial condition values ---
           CLK - clock signal                                  ---
           ENB - input that enables the register               ---
           I   - logic input of n bits                         ---
---Output: O   - logic output of n bits                        ---
-----------------------------------------------------------------*/
module Register #(parameter n=32,parameter init=0)
    (RST,CLK,ENB,I,O);
    input RST,CLK,ENB;
    input [n-1:0] I;
    output reg [n-1:0] O;

    always @ (posedge CLK or posedge RST)
      begin
        if(RST)
```

```
            O <= init;
        else if(ENB)
            O <= I;
        else
            O <= 0;
    end
endmodule
```

3.2.2 Adder and Subtractor

The adder and subtractor are combinational blocks that perform arithmetic operations. In this book, each block is designed as a combinational one and also as a sequential one by including a PIPO register to storage and control the data availability of the result of each operation. The architecture of these blocks is implemented by assigning two input (A,B) and one output (O) buses of n-bit length to perform addition and subtraction. The adder performs the arithmetic operation $O = A + B$, and the subtractor is designed by embedding and adder whose input B is manipulated with 2's complement arithmetic to generate $O = A + (-B)$, which at the end is equivalent to perform $O = A - B$.

The architecture of a combinational block that performs the addition of two digital numbers can be described as shown in Fig. 3.2. It is like a black box that is assigned as *Adder_com* to emphasize that the adder is combinational. Otherwise, if the block description is called *Adder*, without *_com*, in this book such a description is for a sequential adder, as described below. The schematic diagram of a combinational adder is then shown in Fig. 3.2, which has two input busses (A and B) and one output bus (O), each of the same length of bit $n = 32$.

In general, the arithmetic operations must include other pines to process input and output carries. However, as in this book all amplitudes are already known, then the FPGA implementation of chaotic systems and their applications shown in Chap. 7 is synthesized considering computer arithmetic in fixed-point notation. This means that one already knows the maximum (positive) and minimum (negative) values of

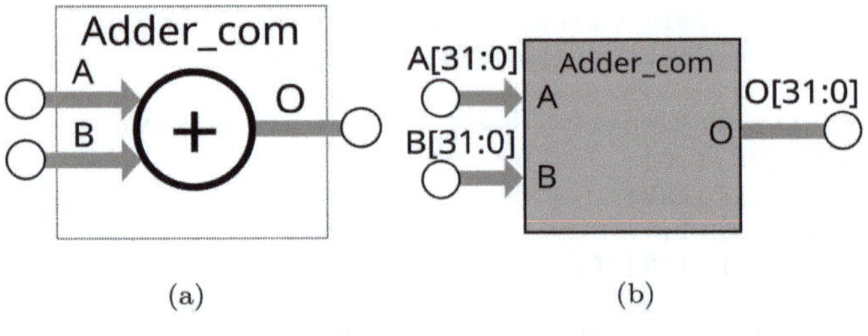

Fig. 3.2 Block representation of a combinational adder: (**a**) internal hardware performing the addition O =A+B and (**b**) description of the buses using $n = 32$ bit length

the state variables of chaotic systems, so that the fixed-point representation includes the use of one bit to represent the sign of a number, and the integer and decimal parts of a number are divided considering the maximum amplitudes, as detailed from Chap. 5. The Verilog description of the combinational adder shown in Fig. 3.2 is given in Listing 3.2.

Listing 3.2 Verilog description of the combinational adder

```
/*-------------------------------------------------------------
---Project: adder of n bits                                ---
---Inputs: A,B are logic signed inputs of n bits         ---
---Output: O produces the arithmetic sum O=A+B           ---
-----------------------------------------------------------*/
module Adder #(parameter n=32)(A,B,O);
    input signed [n-1:0] A, B;
    output [n-1:0] O;

    assign O = A + B;

endmodule
```

As the interest in this book is to control the data processing in a more specific fashion, to increase the operating frequency, then the arithmetic operations such as the addition and subtraction are designed herein to embed a PIPO register. In this manner, the sequential adder and subtractor blocks are designed to have four inputs and one output in the following form: Two input buses are assigned to data in operands A and B, and one output bus is assigned O to provide the arithmetic result, which is controlled by the CLK and RST pines embedded into the PIPO register. Figure 3.3 shows the block design of the sequential adder, whose control is performed through a PIPO register, which has an ENB pin fixed to a high logic 1 state, so that the output bus O can update its current value at each clock cycle (CC).

The Verilog description of the sequential adder shown in Fig. 3.3 is given in Listing 3.3. It can be appreciated that this code instantiates the PIPO register to update the Verilog code of the combinational adder given in Listing 3.2, where

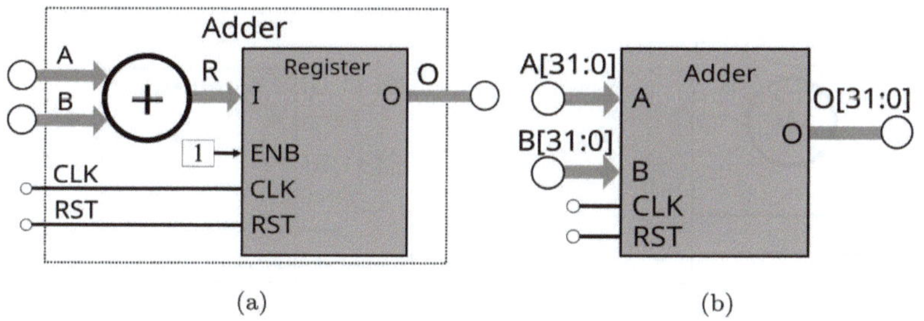

Fig. 3.3 Block representation of a sequential adder: (**a**) internal hardware performing the addition R=A+B and the output O=R by controlling RST and CLK pines and (**b**) description of the buses using $n = 32$ bit length

parameters #(n,0) establish the size of the bus of the PIPO register and the initial condition *init* required in Listing 3.1.

Listing 3.3 Verilog description of the sequential adder

```
/*-----------------------------------------------------------------
---Project: adder of n bits                                    ---
---Inputs: RST - sets the output to zero                       ---
           CLK - clock signal                                  ---
           A,B - logic signed inputs of n bits                 ---
---Output: O   - produces the arithmetic sum of A+B            ---
-----------------------------------------------------------------*/
module Adder #(parameter n=32)(RST,CLK,A,B,O);
    input RST,CLK;
    input signed [n-1:0] A, B;
    output [n-1:0] O;

    wire signed [n-1:0] R;

    assign R = A + B;

    Register #(n,0) U01 (RST,CLK,1'b1,R,O);

endmodule
```

The subtractor is synthesized by using 2's complement in operand B, to accomplish the arithmetic operation $R = A + (-B)$, which at the end is equivalent to perform $R = A - B$, as a combinational block. Figure 3.4 shows the block representation of a sequential subtractor, who also has a PIPO register, as for the sequential adder.

The Verilog description of the sequential subtractor shown in Fig. 3.4 is given in Listing 3.4, in which a PIPO register is instantiated in the same way as done for the Verilog description of the sequential adder. In the Verilog description of the sequential subtractor, the parameters #(n,0) also establish the size of the bus of the PIPO register and the initial condition *init* required in Listing 3.1.

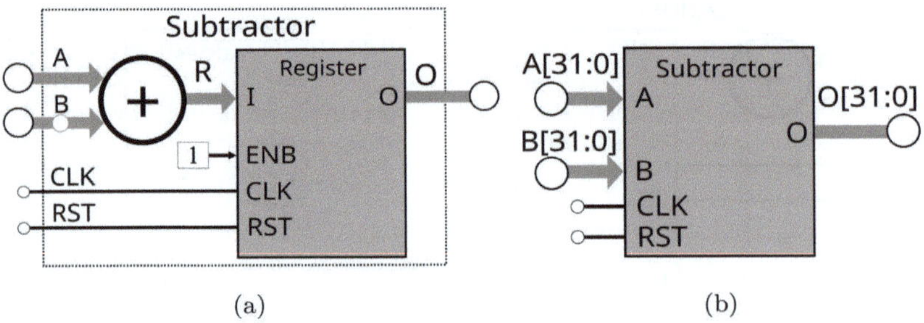

Fig. 3.4 Block representation of a sequential subtractor: (**a**) internal hardware performing the subtraction R=A+(-B) and the output O=R by controlling RST and CLK pines and (**b**) description of the buses using $n = 32$ bit length

Listing 3.4 Verilog description of the sequential subtractor

```
/*-------------------------------------------------------------
---Project: subtractor of n bits                            ---
---Inputs: RST - sets the output to zero                    ---
           CLK - clock signal                               ---
           A,B - logic signed inputs of n bits              ---
---Output: O   - produces the arithmetic subtraction of A-B
    ---
-------------------------------------------------------------*/
module Subtractor #(parameter n=32)(RST,CLK,A,B,O);
    input RST,CLK;
    input signed [n-1:0] A, B;
    output [n-1:0] O;

    wire signed [n-1:0] R;

    assign R = A - B;

    Register #(n,0) U01 (RST,CLK,1'b1,R,O);

endmodule
```

3.2.3 Multiplication Block

The combinational multiplier produces the product between two signed or unsigned binary numbers and allows processing numbers with fixed-point format. The block performs the multiplication between two variables or between one constant and one variable. Its architecture is synthesized in a similar manner as done for the adder and subtractor. If the multiplier to be used is combinational, its name is assigned as *Multiplier_com*; otherwise, if the multiplier is sequential, its block is assigned just as multiplier.

Similar to the block description of the adder and subtractor, a multiplier embeds a combinational *Multiplier_com* and a PIPO register, as shown in Fig. 3.5. Externally, the architecture has two input buses to process the operands A and B of n-bit length has one output bus OT to provide the result of the multiplication, and includes two control pines assigned to CLK and RST. However, as the inputs A and B have size of n bit, and the product OT=A$\times$B produces $2n$ bit, a truncation process must be performed to adapt the output bus OT to have the same size as the input buses, i.e., of n bit that are assigned to the bus R. This data in bus R is further processed to the output of the block, labeled as bus O, by controlling the PIPO register.

In Fig. 3.5a, one can see the truncation block between buses OT and R. The abstract description of the sequential multiplier block has four inputs and one output, as shown in Fig. 3.5b. The Verilog description of the sequential multiplier using numbers in fixed-point format is given in Listing 3.5. In the Verilog code one can see a parameter called *frac*, which defines the size in the decimal part of the number

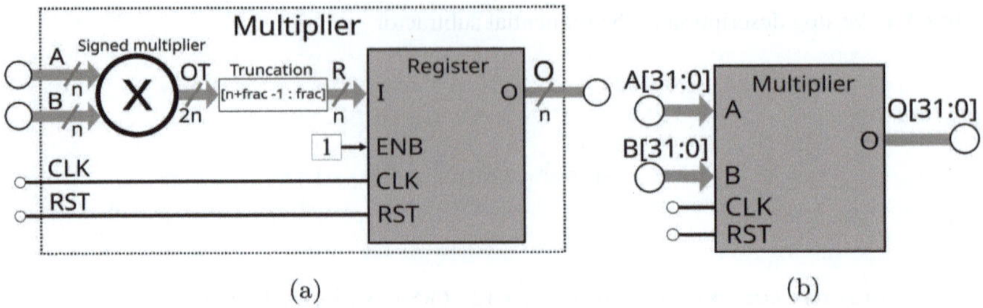

Fig. 3.5 Sequential multiplication block: (**a**) internal diagram including a combinational multiplier block and a PIPO register and (**b**) description of the buses using $n = 32$ bit length

represented in fixed-point format. The buses to process the inputs A and B and the bus to process the product OT = A × B are of signed type, because this multiplier process signed binary numbers. It is worth mentioning that this module can be synthesized into an FPGA depending on the availability of hardware resources, as detailed in [7]. In general, all FPGAs have a certain number of embedded combinational multipliers, but when all of them are used, the synthesis tool can create multiplier blocks using logic elements.

Listing 3.5 Verilog description of the sequential multiplier

```
/*-----------------------------------------------------------------
---Project: multiplier of n bits                                ---
---Inputs: RST - sets the output to zero                        ---
           CLK - clock signal                                   ---
           A,B - logic signed inputs of n bits                  ---
---Output: O   - gives the arithmetic multiplication of A*B    ---
-----------------------------------------------------------------*/
module Multiplier #(parameter n=32,parameter frac=20)
     (RST,CLK,A,B,O);
     localparam msb = n + frac - 1;
     input RST,CLK;
     input signed [n-1:0] A,B;
     output [n-1:0] O;

     wire signed [(2*n)-1:0] OT;
     wire [n-1:0] R;

     assign OT = A * B;

//1.Truncate the result according to the Fixed-point format.
     assign R = OT[msb:frac];

     Register #(n,0) U01 (RST,CLK,1'b1,R,O);

endmodule
```

The description of the sequential multiplier is useful in synthesizing operations processing numbers in fixed-point format, which considers a bit to represent the sign. Otherwise, the Verilog description must avoid the use of the reserved word called *signed*, so that the most significant bit (MSB) must not be considered as the sign of the number. Henceforth, with the goal of avoiding errors when using the sequential multiplier block, the decimal part of a number (*frac*) must always be lower or equal to the number of bits of the multiplier buses for the multiplication operands A and B, for instance.

3.2.4 Description of a Single Constant Multiplier

Some multiplication operations are of the type $a \times x$, where a is a constant number and x is a variable. Operations of this type $a \times x$ can be executed by designing blocks called single constant multiplier (SCM), which is also known as multiplier adder graph (MAG) [8]. An SCM can execute the multiplication by performing addition, subtraction, and shift operations. Basically, the multiplication process consists of executing additions and subtractions of powers of 2 of the variable to be multiplied. The synthesis of this SCM block reduces hardware resources, such as embedded multipliers and logic elements.

The design of an SCM comes from the numerical system with signed-digit (SD) representation, which describes binary numbers using the digits 1, -1, and 0. In this book the SCM is designed for numbers represented with fixed-point format, which requires to represent the constant number with the chosen fixed-point format. Each SCM design is synthesized and described for a specific distribution, so that when it is changed to another distribution of bits, the SCM block will produce wrong results.

One example of designing an SCM is sketched in Fig. 3.6, where the boxes show the operations for the multiplication of type $h \times x$. In this book, h is denoting the step size associated to a numerical method, as shown in Chap. 2, and it appears during the process of discretizing ODEs and FODEs. If $h = 0.01$ and by using fixed-point format 1:11:20 (using 32 bits), the value of h can be represented with the number 10,486, so that the SCM will process $10{,}486 \times x$. In Fig. 3.6, the blocks of type llk imply k shifts to the left, and the other blocks imply addition (+) and subtraction ($-$) operations.

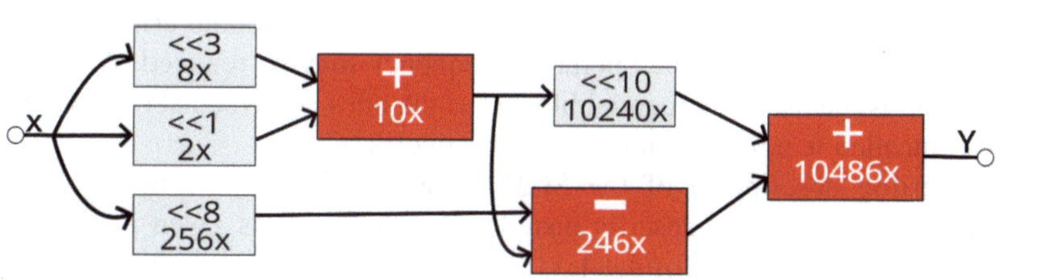

Fig. 3.6 Sketching the SCM block performing $h \times x$, with $h = 0.01$ represented in fixed-point format 1:11:20 (using 32 bits)

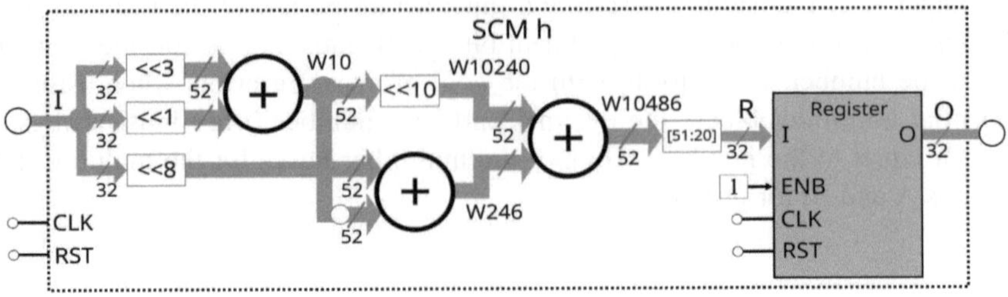

(a)

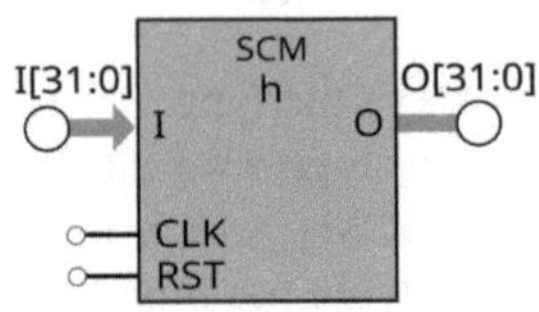

(b)

Fig. 3.7 Hardware design of an SCM for $h = 0.01$: (**a**) internal hardware and (**b**) block representation using 32 bits

The hardware implementation of the SCM shown in Fig. 3.6 is sketched in Fig. 3.7, where the shift and addition operations can be appreciated. Recall that this hardware design is only for $h = 0.01$, so that there is a specific design for different constant values. The architecture consists of 3 inputs and one output designed as follows: one bus of n bit for the input (I), one bus of n bit for the output (O), and two pines for CLK and RST. As one sees, the data I of 32 bits is augmented by 20 bits to have 52 bits to support the three shifts $ll3$, $ll1$, $ll8$. These 52 bits are processed and truncated to 32 at bus R, which is connected at the input of the PIPO register. The truncation operation has the same length of input bits, as the output of the multiplier block, so that the truncation operation is denoted as $[n + frac - 1 : frac]$, where $n = 32$ and *frac*=20.

The design of the SCM shown in Fig. 3.7 uses combinational adders, but as a whole the SCM is sequential, whose data is available by controlling the PIPO register, in which the pin ENB is set to a high value 1, and the CLK and RST pines are controlled externally. A disadvantage of using combinational arithmetic blocks in series connection is that the operation frequency may be reduced. A solution to this problem is the use of sequential blocks, as the ones described above, which data can be controlled in a lower number of clock cycles (CCs).

The Verilog description of the SCM shown in Fig. 3.7 is more elaborated compared to the previous descriptions. The Verilog code is given in Listing 3.6, using 32 bits in fixed-point format 1:11:20. The buses are described as wire signed, and the PIPO is instantiated as Register #(32,0). It can be appreciated that in this type of block description, there is no need to use the reserved word *parameter*, because each SCM architecture is different, and therefore, they are designed according to a specific arithmetic and number of bits.

Listing 3.6 Verilog description of the SCM shown in Fig. 3.7, with $h = 0.01$

```
/*---------------------------------------------------------------
---Project: Single Constant Multiplier h=0.01                ---
---SCM for Fixed-point format: 1:11:20                       ---
---Inputs: RST - sets the output to zero                     ---
           CLK - clock signal                                ---
           I   - logic signed 32-bit input                   ---
---Output: O   - logic 32-bit output                         ---
---------------------------------------------------------------*/
module SCM_h (RST,CLK,I,O);
    input RST,CLK;
    input signed [31:0] I;
    output [31:0] O;

    wire signed [51:0] W10,W246;
    wire signed [51:0] W10240,W10486;
    wire [31:0] R;

    assign W10    = (I<<3) + (I<<1);
    assign W10240 = (W10<<10);
    assign W246   = (I<<8) - W10;
    assign W10486 = W10240 + W246;
    assign R = W10486[51:20];

    Register #(32,0) U01 (RST,CLK,1'b1,R,O);

endmodule
```

3.3 Verilog Descriptions of Other Combinational and Sequential Blocks

The Verilog description of a chaotic system whose ODEs or FODEs are discretized by numerical methods requires the use of the arithmetic blocks described above and other combinational and sequential blocks to control the iterations for solving difference equations. Other combinational and sequential blocks as flip-flops and memories are described in the following subsections.

3.3.1 Multiplexer

The block description of a multiplexer includes pines to assign M control inputs, 2^M data inputs, and one output bus. The inputs and outputs can be parameterized to have a length of n bits. The schematic diagram shown in Fig. 3.8 details a multiplexer having 8 data inputs and one output bus. The Verilog description of this multiplexer is given in Listing 3.7.

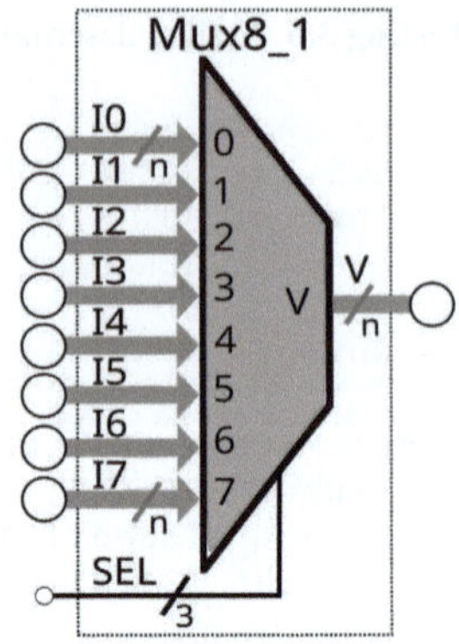

Fig. 3.8 Block description of a multiplexer having 8 inputs and 1 output bus

Listing 3.7 Verilog description of a multiplexer having 8 inputs and one output: Mux8_1

```
/*-------------------------------------------------------------
---Project: 8-to-1 multiplexer with n-bit data lines       ---
---Inputs: SEL - 3-bit control input                       ---
           I0-I7 - n-bit logic inputs                      ---
---Output: O    - n-bit logic output                       ---
-------------------------------------------------------------*/
module Mux8_1 #(parameter n=32) (SEL,I0,I1,I2,I3,I4,I5,I6,I7,V);
    input [2:0] SEL;
    input [n-1:0] I0,I1,I2,I3,I4,I5,I6,I7;
    output [n-1:0] V;

    assign V = (SEL == 3'd0) ? I0:
               (SEL == 3'd1) ? I1:
               (SEL == 3'd2) ? I2:
               (SEL == 3'd3) ? I3:
               (SEL == 3'd4) ? I4:
               (SEL == 3'd5) ? I5:
               (SEL == 3'd6) ? I6:
                               I7;
endmodule
```

3.3.2 Decoder

A decoder is a combinational block having M inputs and 2^M outputs. This block activates one output bus that is controlled by an input in binary code. The architecture of this decoder is shown in Fig. 3.9, who has an input of 3 bits to activate one of the $2^3 = 8$ outputs. Its Verilog description is given in Listing 3.8.

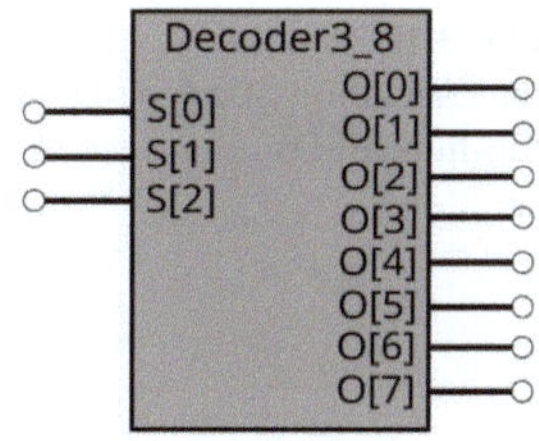

Fig. 3.9 Block representation of a decoder with 3 inputs and 8 outputs

Listing 3.8 Verilog description of a decoder with 3 inputs and 8 outputs: Decoder3_8

```
/*-----------------------------------------------------------------
---Project: 3-to-8 decoder                                     ---
---Inputs: S - 3-bit control input                             ---
---Output: O - 8-bit logic output                              ---
-----------------------------------------------------------------*/
module Decoder3_8 (S,O);
    input [2:0] S;
    output reg [7:0] O;

    always @(*)
     begin
       case(S)
         3'd0: O = 8'h01;
         3'd1: O = 8'h02;
         3'd2: O = 8'h04;
         3'd3: O = 8'h08;
         3'd4: O = 8'h10;
         3'd5: O = 8'h20;
         3'd6: O = 8'h40;
         3'd7: O = 8'h80;
         default: O = 8'h00;
        endcase
     end

endmodule
```

3.3.3 *Type-D Flip-Flop with Enable Pin*

Type-D flip-flop is a sequential circuit with the capacity to hold or retain a logic state 0 or 1. This circuit can change its logic state at each transition of the CLK signal; it can be in the raising or falling edge. Basically, when enable (ENB) pin is in a high logic state 1, the value at the input (D) is charged to the output (Q), at each CLK transition [9].

This type-D flip-flop was used to describe the PIPO register in Sect. 3.2.1, so that it is quite useful to synthesize circuits that require to hold or retain logic values that are controlled by raising or falling CLK edges. Other useful applications of type-

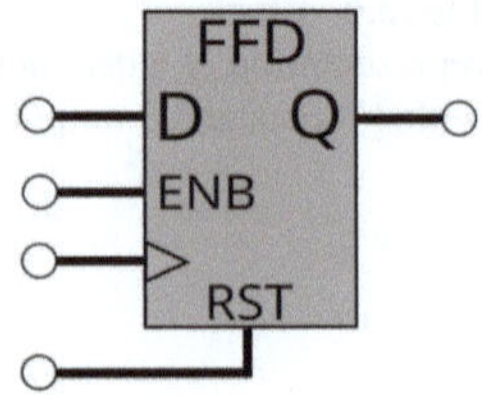

Fig. 3.10 Block representation of a type-D flip-flop (FFD)

D flip-flops include the synthesis of shift registers, counters, finite-state machines (FSMS), and memories, among others.

The block description of the type-D flip-flop (FFD) is shown in Fig. 3.10, which includes an enable (ENB), a CLK, and a reset (RST) pin. The Verilog description is given in Listing 3.9, whose 5 external pines are detailed, and it does not require to be parameterized because it drives just one bit.

Listing 3.9 FFD Verilog code

```
/*----------------------------------------------------------------
---Project: D Flip-Flop with enable pin                       ---
---Inputs: RST - sets the output with 0                       ---
           CLK - clock signal                                 ---
           ENB - input that enables the Flip-Flop             ---
           D   - logic input                                  ---
---Output: Q   - logic output                                 ---
----------------------------------------------------------------*/
module FFD (RST,CLK,ENB,D,Q);
    input RST,CLK,ENB,D;
    output reg Q;

    always @ (posedge CLK or posedge RST)
      begin
        if(RST)
            Q <= 1'b0;
        else if(ENB)
            Q <= D;
        else
            Q <= Q;
      end

endmodule
```

3.3.4 *Ascending Binary Counter*

A binary counter is a sequential circuit that finds a huge number of applications in digital systems. Some examples of applications include its use to count iterations and clock cycles (CCs) and provide access to memory locations, among others. As the blocks include CLK pines, the counter updates its values at each CC. The most

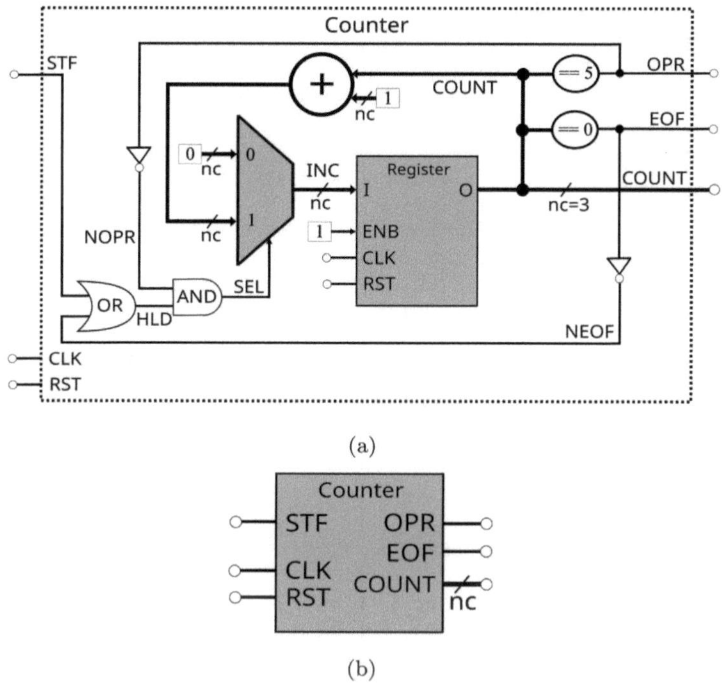

Fig. 3.11 Block description of the ascending counter: (**a**) internal hardware details and (**b**) block description using three inputs and three outputs

common counter designs include a PIPO register, a multiplexer, and an adder [10]. Depending on the design of control inputs, a counter can perform counting in an ascending, descending, or bidirectional fashion.

As an example, Fig. 3.11 shows the internal hardware of the ascending counter of nc bits. The other pines are assigned as follows: The block includes three controls labeled as start counter (STF), CLK, and RST (sets the counter to zero) and has three outputs labeled as COUNT (provides the count value), OPR (gives a logic 1 when the count reaches $nckl$-1), and EOF (gives a logic 1 while COUNT is 0). The architecture begins counting when STF has a high logic value 1. Afterwards, the bus COUNT increments the binary number without stopping at each CC (even if the logic value 1 in STF pin is changed), the counting finished once the value assigned to the output OPR is reached. In this manner, when OPR is activated, the counting returns to zero and waits again for the activation signal at input STF, to begin a new counting process. On the other side, the output EOF remains active while COUNT is in logic value 0. In the internal hardware shown in Fig. 3.11, one can observe the basic logic gates that are in charge of controlling a multiplexer, to decide between increasing the counting value or returning to zero.

The Verilog description of the ascending counter sketched in Fig. 3.11 is given in Listing 3.10. In this architecture, two values are parameterized to establish the number of bits for the counting nc and the maximum number or binary combination to be reached $nclk$.

Listing 3.10 Verilog description of the ascending counter shown in Fig. 3.11

```
/*-----------------------------------------------------------------
---Project: ascending counter 0 to nclk-1                       ---
---Inputs: RST - sets the counter to zero                       ---
           CLK - clock signal                                   ---
           STF - start counter                                  ---
---Output: COUNT - gives the count value                        ---
           OPR - gives '1' when the count reaches nckl-1        ---
           EOF - gives '1' while COUNT is 0                     ---
-----------------------------------------------------------------*/
module Counter #(parameter nc=3,nclk=6) (RST,CLK,STF,COUNT,OPR,
    EOF);
    input RST,CLK,STF;
    output OPR,EOF;
    output [nc-1:0] COUNT;

    wire [nc-1:0] INC;
    wire NEOF, NOPR, HLD, SEL;

    assign OPR = (COUNT == nclk-1'b1) ? 1'b1 : 1'b0;
    assign EOF = (COUNT == 0) ? 1'b1 : 1'b0;
    assign INC = (SEL) ? COUNT + 1'b1: {nc{1'b0}};

    assign NEOF = ~EOF;
    assign NOPR = ~OPR;
    assign  HLD = STF | NEOF;
    assign  SEL = HLD & NOPR;

    Register #(nc,0) U01 (RST,CLK,1'b1,INC,COUNT);

endmodule
```

3.3.5 *Bidirectional Binary Counter*

The architecture of a bidirectional binary counter is shown in Fig. 3.12, having nc bits that are counting in a bidirectional fashion according to the logic state of the bus control OPC. At each clock cycle (CC), the binary counter updates its value according to the value coming from a multiplexer of 4 inputs and 1 output. The way this bidirectional counter works depends on the value of the control pin OPC, which can select four operation modes driving by the multiplexer4_1: Set the counter to zero when OPC=0, hold the current counting value when OPC=1, perform ascending counting when OPC=2, and perform descending counting when OPC=3.

The Verilog description of the bidirectional counter sketched in Fig. 3.12 is given in Listing 3.11. Similar to the ascending counter, two values are parameterized to

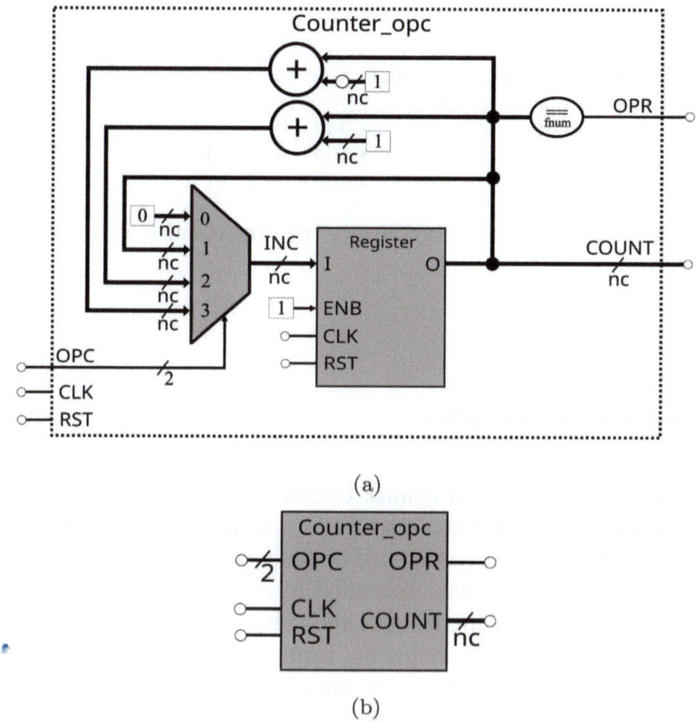

Fig. 3.12 Block description of the bidirectional counter: (**a**) internal hardware details and (**b**) block description using three inputs and two outputs

establish the number of bits for the counting *nc* and the number in the current count *fnum*. It is important that the parameters *nclk* and *fnum* be specified in natural numbers, which are within the range of values defined by *nc*.

Listing 3.11 Verilog description of the bidirectional counter shown in Fig. 3.12

```
/*------------------------------------------------------------------
---Project: bidirectional counter                                ---
---Inputs: RST - sets the counter to zero                        ---
           CLK - clock signal                                    ---
           OPC - input that sets the mode of the counter         ---
---Output: COUNT - gives the count value                         ---
           OPR - gives '1' when count equals fnum                ---
------------------------------------------------------------------*/
module Counter_opc #(parameter nc=3,parameter fnum=6)
    (RST,CLK,OPC,COUNT,OPR);
    input RST,CLK;
    input [1:0] OPC;
    output [nc-1:0] COUNT;
    output OPR;

```

```
    wire [nc-1:0] INC;

    assign OPR = (COUNT == fnum) ? 1'b1 : 1'b0;
    assign INC = (OPC==2'b00) ? {nc{1'b0}}  :
                 (OPC==2'b01) ? COUNT       :
                 (OPC==2'b10) ? COUNT + 1'b1:
                                COUNT - 1'b1;

    Register #(nc,0) U01 (RST,CLK,1'b1,INC,COUNT);

endmodule
```

3.3.6 *Finite-State Machine*

Finite-state machine (FSM) is a control circuit that is synthesized by using flip-flops. The main objective of designing a FSM is to control the clock cycles (CCs) that synchronize the flow of execution of operations in a sequential system, as the synthesis of difference equations associated to chaotic systems. A FSM defines states that can be accessed depending of the values of the control inputs. A FSM can be synchronous or asynchronous, the former being one of the most used. In literature, the FSMs most known were introduced and designed by Moore and another designed by Mealey [11]. Figure 3.13 sketches an example of a transition diagram of a Moore-type machine having four states and in which the outputs of the block only depend of the current state.

The Verilog description of the Moore-type FSM shown in Fig. 3.13 is given in Listing 3.12. The Verilog code is available from a template provided by the software Quartus© II.

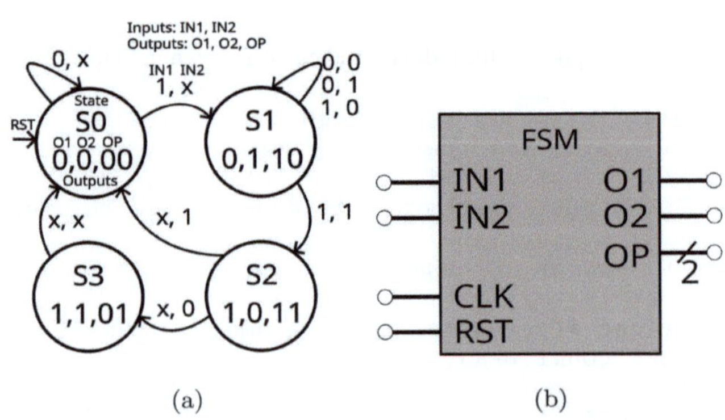

Fig. 3.13 Moore-type FSM: (**a**) transition diagram and (**b**) block description using four inputs (IN1, IN2, CLK, RST) and three outputs (O1, O2, OP)

Listing 3.12 Verilog description of a Moore-type FSM based on Quartus© II template

```
/*------------------------------------------------------------------
---Project: Moore type FSM based on Quartus II Verilog
                  ---
---Inputs: RST - sets the initial state                         ---
           CLK - clock signal                                   ---
           IN1,IN2 - logic inputs                               ---
---Output: O1,O2,OP - logic outputs                             ---
-----------------------------------------------------------------*/
module FSM(RST,CLK,IN1,IN2,O1,O2,OP);
    localparam S0=0, S1=1, S2=2, S3=3;

    input RST,CLK,IN1,IN2;
    output reg O1,O2;
    output reg [1:0] OP;
    reg [1:0] EDO;

    always @ (EDO) begin
        case (EDO)
            S0:
              begin O1<=1'b0; O2<=1'b0; OP<=2'b00; end
            S1:
              begin O1<=1'b0; O2<=1'b1; OP<=2'b10; end
            S2:
              begin O1<=1'b1; O2<=1'b0; OP<=2'b11; end
            S3:
              begin O1<=1'b1; O2<=1'b1; OP<=2'b01; end
            default:
              begin O1<=1'b0; O2<=1'b0; OP<=2'b00; end
        endcase
    end

    always @ (posedge CLK or posedge RST) begin
        if (RST)
            EDO <= S0;
        else
            case (EDO)
                S0:
                  if(IN1)
                    EDO<=S1;
                  else
                    EDO<=S0;
                S1:
                  if(IN1 && IN2)
                    EDO<=S2;
                  else
                    EDO<=S1;
                S2:
                  if(IN2)
                    EDO<=S0;
                  else
                    EDO<=S3;
                S3:
                    EDO<=S0;
```

```
            endcase
    end
endmodule
```

3.3.7 *Random Access Memory (RAM)*

The random access memory (RAM) block is a volatile memory that allows both writing and reading. It is most used to store temporal data. This type of RAM allows access to any memory cell through the control of a memory address port.

Generally, an FPGA board includes RAM matrices, which are specified by their deep and length of data. A RAM has different sizes, and in Listing 3.13 one can see the Verilog description of a synchronous 256×32 RAM with simple port. This code is based on a template provided by software Quartus© II. One can see the use of the parameter reserved word to establish the default length to 32 bits that can generate $2^8 = 256$ address, i.e., the address bus is of 8-bit length.

Listing 3.13 Verilog description of a synchronous 256×32 RAM based on Quartus© II template

```
/*-----------------------------------------------------------------
---Project: RAM memory based on Quartus II                     ---
---Inputs: CLK - clock signal                                  ---
           WE - write enable                                   ---
           DATA - data to store in memory                      ---
           ADDR - memory address                               ---
---Output: Q - show stored data                                ---
---------------------------------------------------------------*/
module RAM #(parameter data_width=32, parameter addr_width=8
    (CLK,WE,DATA,ADDR,Q);
    localparam k = 2**addr_width;

    input CLK,WE;
    input [(data_width-1):0] DATA;
    input [(addr_width-1):0] ADDR;
    output [(data_width-1):0] Q;

    reg [data_width-1:0] RAM_R[k-1:0];
    reg [addr_width-1:0] ADDR_R;

    always @ (posedge CLK)
    begin
        if (WE)
            RAM_R[ADDR] <= DATA;

        ADDR_R <= ADDR;
    end

    assign Q = RAM_R[ADDR_R];

endmodule
```

3.3.8 *Read-Only Memory (ROM)*

The read-only memory (ROM) block is an only read memory. The data is previously stored so that one cannot perform a writing process during an execution time. The RAM matrices of an FPGA can be used to describe the hardware of a ROM. In this manner, the size of a ROM is also specified by its deep and length of data.

The Verilog description of a synchronous ROM of one output port is given in Listing 3.14. This code is based on a template provided by software Quartus© II. The Verilog description of this type of memory requires to add a *.txt file that must contain the ROM data stored. Again, one can see the use of parameter reserved word to establish the default length to 32 bits that can generate $2^6 = 64$ address, i.e., the address bus is of 6-bit length.

Listing 3.14 Verilog description of a synchronous ROM based on Quartus© II template

```
/*-------------------------------------------------------------
---Project: ROM based on Quartus II                        ---
---Inputs: CLK - clock signal                                   ---
           ADDR - memory address                                ---
---Output: Q - show stored data                                 ---
-------------------------------------------------------------*/
module ROM #(parameter data_width=32, parameter addr_width=6)
    (CLK,ADDR,Q);
    localparam k = 2**addr_width;

    input CLK;
    input [(addr_width-1):0] ADDR;
    output reg [(data_width-1):0] Q;

    reg [data_width-1:0] ROM_R[0:k-1];

    initial
    begin
        $readmemb("File.txt", ROM_R);
    end

    always @ (posedge CLK)
    begin
        Q <= ROM_R[ADDR];
    end

endmodule
```

Figure 3.14 shows the block description of the RAM and ROM having data buses of 32-bit length, address bus of 6-bit length to address $2^6 = 64$ cells, and showing the input and output ports.

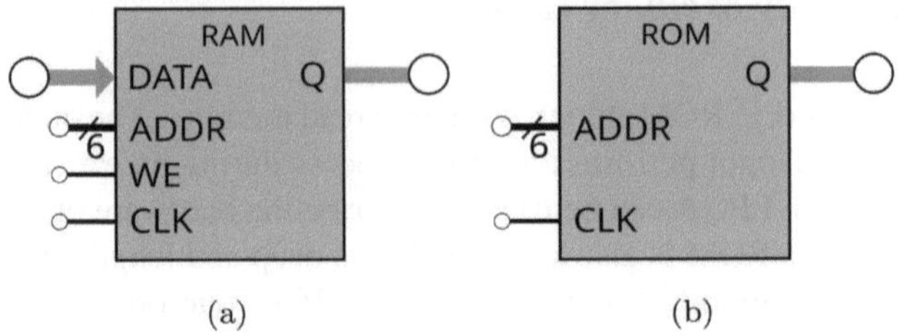

Fig. 3.14 Block description of the (**a**) RAM and (**b**) ROM with data buses of 32 bits

3.3.9 Register-Based Memory

The register-based memory (RBM) is a type of temporal memory that can be synthesized by using parallel input and output or PIPO registers. Its function is similar to a RAM. It uses an address bus to select the cell to store the data. In addition, this RBM includes other operation modes, for example: If the PIPO registers are series connected, the data stored in each register can be shifted, and if the control includes multiplexers, one can select between performing a shift operation with all PIPO registers and defining through the address pointer to which the registered data will be stored. This type of memory is quite useful to synthesize fractional-order chaotic systems, as shown in Chap. 6, where the FPGA implementation is based on the short-memory principle of a numerical algorithm, which requires to shift all data once all memory cells are filled.

The block description of the RBM is shown in Fig. 3.15. One can see details of the PIPO registers controlled by multiplexers, a decoder, and logic gates. This example is for the synthesis of an RBM of size 8×32.

The Verilog description of the RBM of size 8×32 is given in Listing 3.15. The use of the reserved word parameter to establish the length of the data bus to $n = 32$ bit can be seen. This parameter n can then be varied during the Verilog programming. A disadvantage of this type of memory is that for a huge number of registers into a module, for example, for an RBM of size 512×32, the hardware consumption of logic elements in an FPGA board can be considerably high.

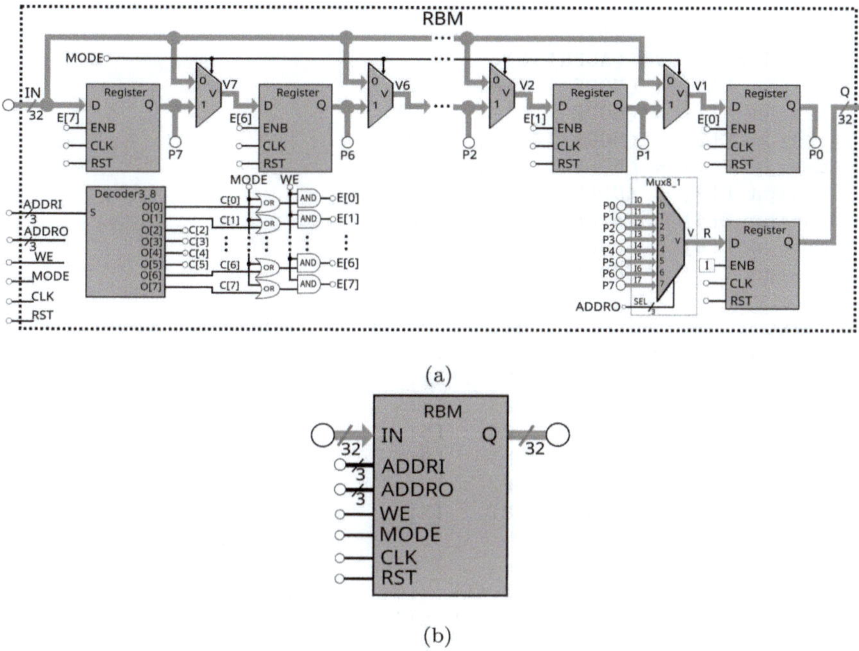

Fig. 3.15 Block description of the register-based memory (RBM) with a size 8×32): (**a**) internal hardware description and (**b**) block description with 7 inputs and one output

Listing 3.15 Verilog description of an RBM

```
/*------------------------------------------------------------
---Project: Register-based memory 8x32                     ---
---Inputs: RST - sets all registers to zero                ---
           CLK - clock signal                              ---
           WE - write enable                               ---
           IN - data to store in memory                    ---
           ADDRI - memory address to save                  ---
           ADDRO - memory address to get a value           ---
---Output: Q - value from a register                       ---
------------------------------------------------------------*/
module RBM #(parameter n=32) (RST,CLK,WE,MODE,IN,ADDRI,ADDRO,Q);
    localparam naddr = 3;
    localparam nenb = 2**naddr;

    input RST,CLK,WE,MODE;
    input [n-1:0] IN;
    input [naddr-1:0] ADDRI,ADDRO;
    output [n-1:0] Q;

    wire [n-1:0] R;
    wire [n-1:0] V7,V6,V5,V4,V3,V2,V1;
    wire [n-1:0] P7,P6,P5,P4,P3,P2,P1,P0;
    wire [nenb-1:0] C, E;
```

```

//1.Decode which register stores data.
    Decoder3_8 U00 (ADDRI,C);
    assign E[0] = (MODE | C[0]) & WE;
    assign E[1] = (MODE | C[1]) & WE;
    assign E[2] = (MODE | C[2]) & WE;
    assign E[3] = (MODE | C[3]) & WE;
    assign E[4] = (MODE | C[4]) & WE;
    assign E[5] = (MODE | C[5]) & WE;
    assign E[6] = (MODE | C[6]) & WE;
    assign E[7] = (MODE | C[7]) & WE;

//2.Muxes to store in a register or shift each register
    assign  V7 = (MODE) ? P7 : IN;
    assign  V6 = (MODE) ? P6 : IN;
    assign  V5 = (MODE) ? P5 : IN;
    assign  V4 = (MODE) ? P4 : IN;
    assign  V3 = (MODE) ? P3 : IN;
    assign  V2 = (MODE) ? P2 : IN;
    assign  V1 = (MODE) ? P1 : IN;

//3.Registers
    Register #(n,0) U01 (RST,CLK,E[0],V1,P0);
    Register #(n,0) U02 (RST,CLK,E[1],V2,P1);
    Register #(n,0) U03 (RST,CLK,E[2],V3,P2);
    Register #(n,0) U04 (RST,CLK,E[3],V4,P3);
    Register #(n,0) U05 (RST,CLK,E[4],V5,P4);
    Register #(n,0) U06 (RST,CLK,E[5],V6,P5);
    Register #(n,0) U07 (RST,CLK,E[6],V7,P6);
    Register #(n,0) U08 (RST,CLK,E[7],IN,P7);

//4.Memory output
    Mux8_1 #(n) U09 (ADDRO,P0,P1,P2,P3,P4,P5,P6,P7,R);
    Register #(n,0) U10 (RST,CLK,1'b1,R,Q);

endmodule
```

If one is describing an RBM with a high number of memory cells, the Verilog code will include more register instantiations, and the size of the output multiplexer and input decoder must be changed. The Verilog description is repetitive, but this process of augmenting the size of the blocks is appropriate to synthesize an RBM with one shift mode.

3.3.10 *Multiplier Accumulator*

A very useful block for FPGA implementations is composed of a multiplier and an accumulator (MAC) [8], also known as accumulative sum [2]. Basically, a MAC consists of a multiplier and adder and two PIPO registers. The MAC block allows accumulating the result of a sum of products between the inputs labeled as V and

K. Although the MAC block requires of a higher number of clock cycles (CCs) to provide a final result, its hardware design saves resources through performing multiplications and successive sums with a single block. The iterative process in a MAC requires of a control unit that can be designed with a FSM and a counter to establish the number of sum of products, to store the final result, and to delete the accumulated value.

The block description of the MAC block is shown in Fig. 3.16. It can be appreciated that this block has two input buses labeled as V and K that are assigned to process the input numbers being multiplied and then, the bus labeled S provides the final result. This block has five control pines that perform the following operations: The pin assigned to load the accumulator (LAC) enables the multiplier and the accumulator register. The pin assigned to load the final register (LFR) activates the register that stores the final accumulated value. The pin assigned to clear the accumulator (CAC) deletes the value that is stored in the accumulator register. The pines CLK and RST perform the operations to drive the clock signal and reset the block, as done in the previous blocks detailed above. As shown in

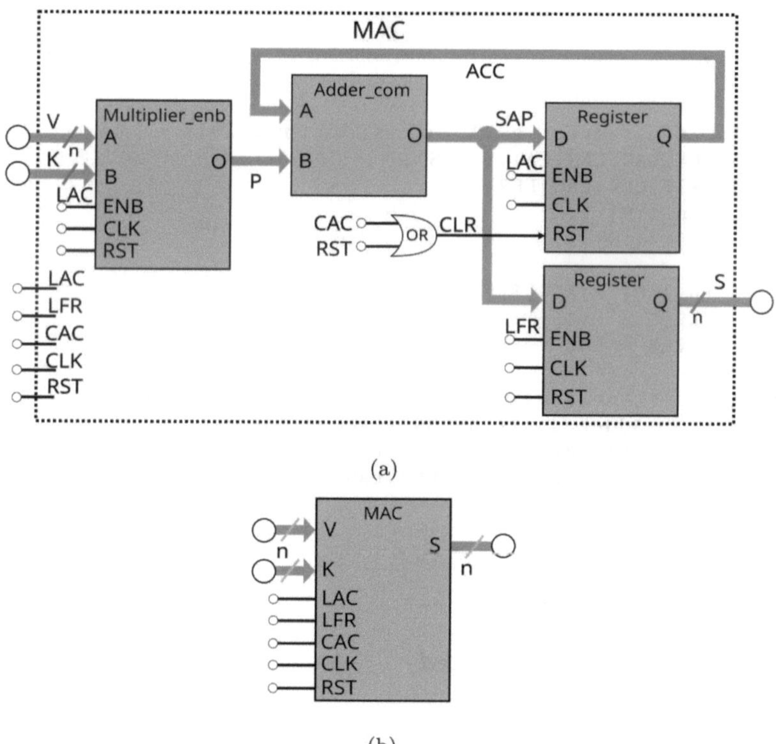

Fig. 3.16 Block description of the multiplier accumulator (MAC): (**a**) internal hardware description and (**b**) block representation consisting of seven inputs and one output

Fig. 3.16a, the description of this module includes a block labeled Multiplier_enb, who is a multiplier with an enable pin to control the beginning of the multiplication process. One can find other FPGA implementations that include the combinational block labeled Multiplier_com; however, the selection of the blocks will depend of the design of the FSM for the MAC block.

The Verilog description of the multiplier accumulator block is given in Listing 3.16. As shown in Fig. 3.16, it uses a combinational adder connected to a PIPO register. As the variables being multiplied have a bus of n bit, the multiplication results can be twice. In this manner, the Verilog description includes the parameterization of the n bit and the decimal part is increased by *frac*. This process is instantiated in the instruction Multiplier_enb #(n,frac).

Listing 3.16 Verilog description of the multiplier accumulator shown in Fig. 3.16

```
/*-----------------------------------------------------------------
---Project: Multiplier-Accumulator of n bits                     ---
---Inputs: RST - sets the output to zero                         ---
           CLK - clock signal                                    ---
           LAC - load accumulator register                       ---
           LFR - load final result                               ---
           CAC - clear accumulator register                      ---
           V,K - logic signed inputs of n bits                   ---
---Output: S   - gives the arithmetic sum of products            ---
---------------------------------------------------------------*/
module MAC #(parameter n=32, parameter frac=20)
    (RST,CLK,LAC,LFR,CAC,V,K,S);
    input RST,CLK,LAC,LFR,CAC;
    input [n-1:0] V,K;
    output [n-1:0] S;

    wire [n-1:0] P,ACC,SAP;
    wire CLR;
    assign CLR = CAC | RST;

    Multiplier_enb #(n,frac) U01 (RST,CLK,LAC,V,K,P);
             Adder_com #(n) U02 (ACC,P,SAP);
            Register #(n,0) U03 (CLR,CLK,LAC,SAP,ACC);
            Register #(n,0) U04 (RST,CLK,LFR,SAP,S);

endmodule
```

3.3.11 Multiplier-Multiplexed

The previous blocks are quite useful to synthesize chaotic systems into an FPGA. Other combinations of basic blocks can also be described to improve the hardware design of chaotic systems. The combination of a multiplier with a multiplexer is useful to square two state variables and get their product. This process can be performed by reusing the same multiplier block. Another advantage of the

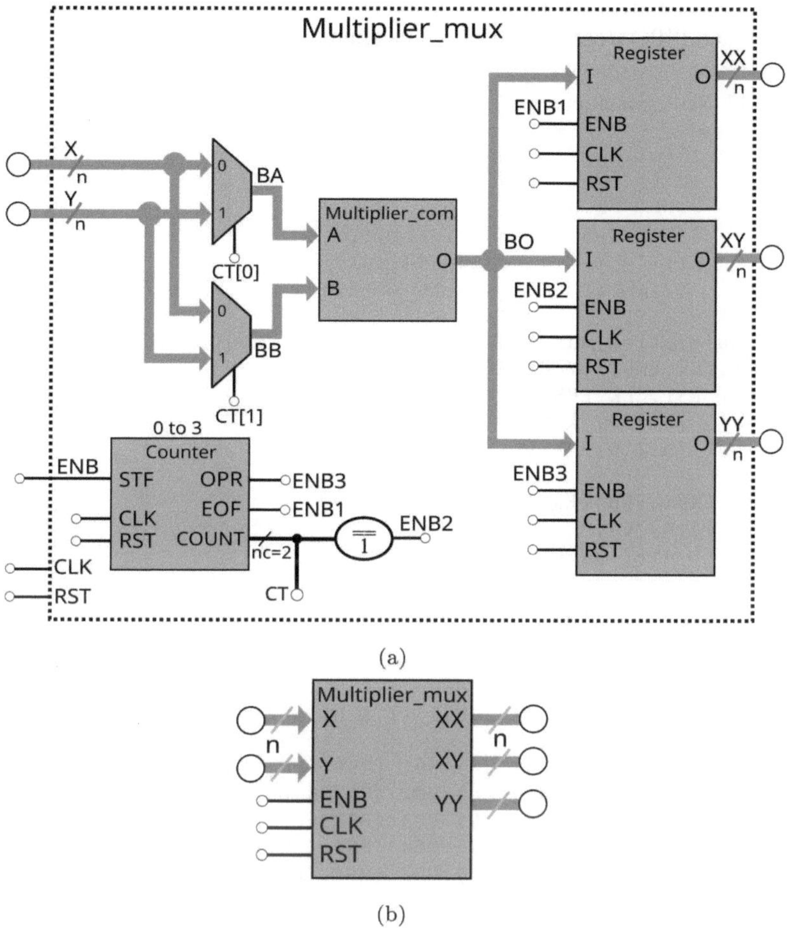

Fig. 3.17 Block description of a multiplier-multiplexed: (**a**) internal hardware details and (**b**) description as a black box

multiplier-multiplexed is the saving of hardware resources in architectures with a huge number of bits. The design of the multiplier-multiplexed is shown in Fig. 3.17, and it consists of a combinational multiplier, three PIPO registers, two multiplexers, and one counter. The multiplier-multiplexed requires four clock cycles (CCs) to produce a result. The counter is in charge of changing the inputs in the multiplier and selecting one of the three PIPO registers through an enable (ENB) pin and is in charge of storing the result of each product in a single PIPO register. Figure 3.17b shows the block of the multiplier-multiplexed with two input buses of n data, three busses of n output data, and three control pines, namely, ENB, CLK, and RST.

The Verilog description of the multiplier-multiplexed Multiplier_mux block shown in Fig. 3.17 is given in Listing 3.17, where the parameterization of the word data length n and the instantiation of Counter #(2,4), combinational Multiplier_com #(n,frac), and PIPO Register #(n,0) can be appreciated.

Listing 3.17 Verilog description of a multiplier-multiplexed Multiplier_mux block

```
/*-------------------------------------------------------------------
---Project: Multiplexed multiplier of n bits                     ---
---Inputs: RST - sets the output to zero                         ---
           CLK - clock signal                                    ---
           ENB - enable multiplications                          ---
           X,Y - n-bit inputs to multiply                        ---
---Output: XX,XY,XY  - products obtained by multiplication        ---
-------------------------------------------------------------------*/
module Multiplier_mux #(parameter n=32,parameter frac=20)
    (RST,CLK,ENB,X,Y,XX,XY,YY);
    input RST,CLK,ENB;
    input  [n-1:0] X,Y;
    output [n-1:0] XX,XY,YY;

    wire ENB1,ENB2,ENB3;
    wire [1:0] CT;
    wire [n-1:0] BA,BB,BO;

    assign BA = (CT[0]) ? Y:X;
    assign BB = (CT[1]) ? Y:X;

    assign ENB1 = (CT == 2'b00) ? 1'b1 : 1'b0;
    assign ENB2 = (CT == 2'b01) ? 1'b1 : 1'b0;

              Counter #(2,4) U01 (RST,CLK,ENB,CT,ENB3);
    Multiplier_com #(n,frac) U02 (BA,BB,BO);
             Register #(n,0) U03 (RST,CLK,ENB1,BO,XX);
             Register #(n,0) U04 (RST,CLK,ENB2,BO,XY);
             Register #(n,0) U05 (RST,CLK,ENB3,BO,YY);

endmodule
```

The blocks described above are enough to synthesize the chaotic systems described in Chap. 2. However, there exist other basic blocks such as encoders, demultiplexers, type-JK flip-flops, type-T flip-flops, shift registers, and arithmetic dividers, among others, that are not described in this book.

3.4 Verilog Description of a 2D Chaotic Map and 3D Lorenz System

This section shows the use of the Verilog descriptions given above to synthesize the 3D Lorenz system and a 2D chaotic map.

3.4.1 Verilog Description of the 3D Lorenz System

The first case study is the well-known 3D Lorenz system whose mathematical model is given in Chap. 2 in Eq. (2.6). It is a continuous-time system modeled by three ODEs, which can be solved by applying a numerical method. By applying the forward Euler (FE) method, the discretization of the ODEs leads us to the difference equations given in (3.1), where h is the step size. As one can infer, these equations can be solved by designing digital blocks associated to multipliers, adders, and subtractors. The multiplication of the state variables by h, σ, ρ, β can be performed by using single constant multiplier (SCM) blocks, as described in Sect. 3.2.4.

$$\begin{aligned} x[n+1] &= x[n] + h(\sigma(y[n] - x[n])) \\ y[n+1] &= y[n] + h(x[n](\rho - z[n])) - y[n]) \\ z[n+1] &= z[n] + h(x[n]y[n] - \beta z[n]) \end{aligned} \tag{3.1}$$

The discretized equations of Lorenz system can generate chaotic behavior by setting $\sigma = 10$, $\rho = 28$, $\beta = 8/3$, and $h = 0.01$.

The Verilog description of (3.1) requires arithmetic operations; however, the length of bits to describe each adder, subtractor, and multiplier block depends on a previous analysis of the amplitudes of the state variables and of the products in the equations. In addition to the analysis of the amplitudes, the computer arithmetic should be defined to identify the architecture of the arithmetic blocks. In this book, the representation of the numbers is performed by using fixed-point notation.

3.4.2 Fixed-Point Notation for Lorenz System

To establish the computer arithmetic for the design of the building blocks, the first step is the simulation of the mathematical model to find the maximum number among all operations and, for chaotic systems, by using different initial conditions to solve ODEs or FODEs. This stage helps to determine the maximum and minimum amplitude values to establish the number of bits in the integer and decimal (*frac*) parts of the fixed-point representation. Simulating the Lorenz system, the maximum and minimum amplitudes are within the range ± 2048. This number can be represented using 11 bits ($\pm 2^{11} = 2048$). The sign can be represented by adding one bit as the most significant bit (MSB), so that by using 32 bits, the fixed-point notation can be expressed as 1:11:20, meaning that the MSB is used to represent the sign, the 11 bits the integer part, and the remaining 20 bits representing the decimal part of real numbers coming from the simulation of the mathematical models discretized by numerical methods. This distribution for the Lorenz system using 32 bits can be represented as shown in Table 3.1.

Table 3.1 Fixed-point notation using 32 bits for the Lorenz system

Sign	Integer part	Decimal part
1 bit	11 bits	20 bits
0	00000000000	.00000000000000000000

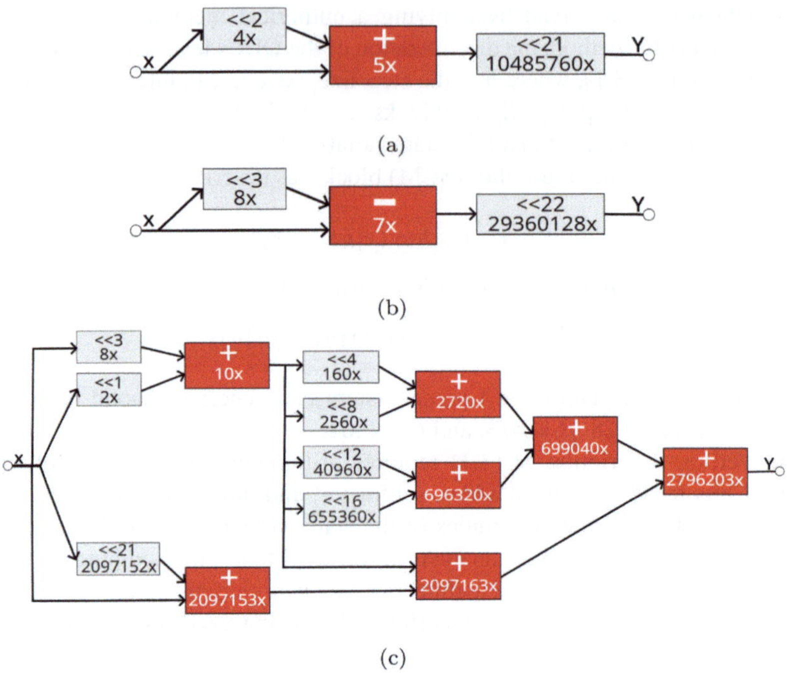

Fig. 3.18 Shift and addition/subtraction operations for designing SCMs with 32 bits in fixed-point notation, associated to the constants: (**a**) σ, (**b**) ρ, and (**c**) β

3.4.3 Block and Verilog Descriptions of the Constants Multiplying Variables for Lorenz System

After choosing the fixed-point distribution, one can identify the blocks to be designed using 32 bits for computer arithmetic. The block description of discretized equations, as the ones given in Eq. (3.1), can be synthesized in different ways. However, one can associate multiplications between two variables and multiplications between one constant and one variable, where hardware resources can be saved by designing SCMs. In all cases or design styles, the goal is devoted to increase the operating frequency along low hardware resource consumption. In Eq. (3.1), four constants are appreciated, h, σ, ρ, β, and they are multiplying state variables, so that SCMs can be synthesized in all these cases. The shift and addition/subtraction operations for designing SCMs associated to σ, ρ, β are shown in Fig. 3.18, which use the distribution of bits listed in Table 3.1.

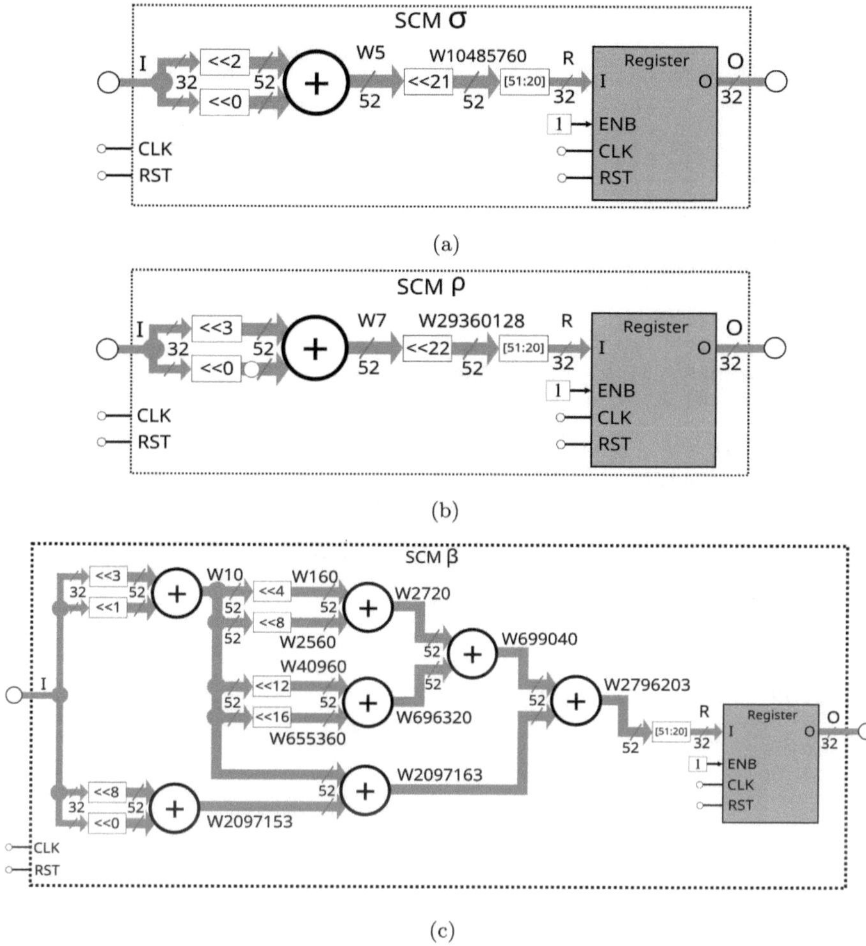

Fig. 3.19 Hardware design of the flow diagrams shown in Fig. 3.18, to synthesize the SCMs for the constant: (**a**) σ, (**b**) ρ, and (**c**) β

The internal hardware designs of the flow diagrams shown in Fig. 3.18, are detailed in Fig. 3.19. For the case of the step size h, the shift and arithmetic operations are sketched in Fig. 3.6, and its internal hardware design is shown in Fig. 3.7. All blocks are designed by using the same fixed-point representation of 32 bit.

The Verilog description for the synthesis of the SCMs shown in Fig. 3.19, are given in Listing 3.18 for the constant σ, Listing 3.19 for ρ, and Listing 3.20 for β. Recall that in all cases for SCM design, a PIPO register is instantiated, such as Register #(32,0).

Listing 3.18 Verilog description of the SCM for $\sigma = 10$

```
/*-------------------------------------------------------------
---Project: Single Constant Multiplier sigma s=10          ---
---SCM for Fixed-point format: 1:11:20                     ---
---Inputs: RST - sets the output to zero                   ---
           CLK - clock signal                              ---
           I   - logic signed 32-bit input                 ---
---Output: O   - logic 32-bit output                       ---
-------------------------------------------------------------*/
module SCM_10 (RST,CLK,I,O);
    input RST,CLK;
    input signed [31:0] I;
    output [31:0] O;

    wire signed [51:0] W5, W10485760;
    wire [31:0] R;

    assign W5        = (I<<2) + I;
    assign W10485760 = (W5<<21);
    assign R         = W10485760[51:20];

    Register #(32,0) U01 (RST,CLK,1'b1,R,O);

endmodule
```

Listing 3.19 Verilog description of the SCM for $\rho = 28$

```
/*-------------------------------------------------------------
---Project: Single Constant Multiplier rho r=28            ---
---SCM for Fixed-point format: 1:11:20                     ---
---Inputs: RST - sets the output to zero                   ---
           CLK - clock signal                              ---
           I   - logic signed 32-bit input                 ---
---Output: O   - logic 32-bit output                       ---
-------------------------------------------------------------*/

module SCM_28 (RST,CLK,I,O);
    input RST,CLK;
    input signed [31:0] I;
    output [31:0] O;

    wire signed [51:0] W7, W29360128;
    wire [31:0] R;

    assign W7        = (I<<3) - I;
    assign W29360128 = (W7<<22);
    assign R         = W29360128[51:20];

    Register #(32,0) U01 (RST,CLK,1'b1,R,O);

endmodule
```

Listing 3.20 Verilog description of the SCM for $\beta = 8/3$

```
/*-------------------------------------------------------------
---Project: Single Constant Multiplier beta b=8/3          ---
---SCM for Fixed-point format: 1:11:20                     ---
---Inputs: RST - sets the output to zero                   ---
           CLK - clock signal                              ---
           I   - logic signed 32-bit input                 ---
---Output: O   - logic 32-bit output                       ---
-------------------------------------------------------------*/
module SCM_8_3 (RST,CLK,I,O);
    input RST,CLK;
    input signed [31:0] I;
    output [31:0] O;

    wire signed [51:0] W10,W160,W2560,W2720,W40960;
    wire signed [51:0] W655360,W696320,W699040;
    wire signed [51:0] W2097153,W2097163,W2796203;
    wire [31:0] R;

    assign W10      = (I<<3)+(I<<1);
    assign W2097153 = (I<<21) + I;

    assign W160     = (W10<<4);
    assign W2560    = (W10<<8);
    assign W40960   = (W10<<12);
    assign W655360  = (W10<<16);

    assign W2720    = W160 + W2560;
    assign W696320  = W40960 + W655360;
    assign W2097163 = W10 + W2097153;

    assign W699040  = W2720 + W696320;
    assign W2796203 = W699040 + W2097163;

    assign R        = W2796203[51:20];

    Register #(32,0) U01 (RST,CLK,1'b1,R,O);

endmodule
```

In some cases, the discretized equations can be factorized to perform repetitive evaluations of sets of arithmetic operations. Observing the numerical method forward Euler (FE), the repetitive evaluation of operations can be identified, which can be performed by multiplying the constant h and one addition operation with a state variable, such as $x[n] + h(...)$. This can be instantiated by a block to perform the integration and it can be labeled as Integrator_FE, as shown in Fig. 3.20, where the use of an SCM to represent the multiplication of a state variable by h and a sequential adder can be appreciated, so that in both blocks a CLK and RST pines are required.

The Verilog description of the blocks shown in Fig. 3.20 is given in Listing 3.21. The instantiation of the two main subblocks labeled as SCM_h and Adder #(32) can be appreciated.

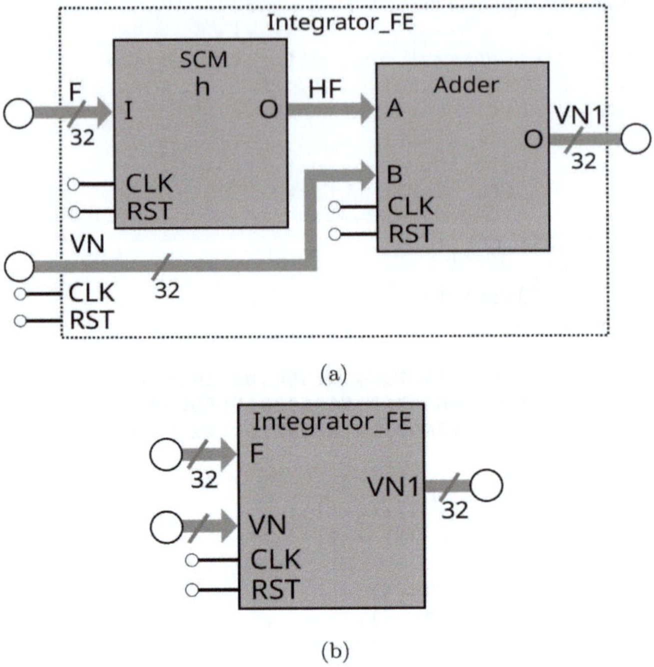

Fig. 3.20 Block description of the integrator by applying forward Euler method: (**a**) internal hardware description and (**b**) black box representation

Listing 3.21 Verilog description of the block Integrator_FE shown in Fig. 3.20

```
/*------------------------------------------------------------------
---Project: Integrator FE Block                                  ---
---Inputs: RST  - sets the output to zero                        ---
           CLK  - clock signal                                   ---
           F,VN - logic 32-bit inputs                            ---
---Output: VN1  - logic 32-bit output                            ---
---------------------------------------------------------------*/
module Integrator_FE_Block (RST,CLK,F,VN,VN1);
     input RST,CLK;
     input [31:0] F,VN;
     output [31:0] VN1;

     wire [31:0] HF,VF;

          SCM_h U01 (RST,CLK,F,HF);
    Adder #(32) U02 (RST,CLK,HF,VN,VN1);

endmodule
```

One can infer that this integrator block can be changed by the numerical method being applied and that the SCM for h must also be updated for each value of the constant. The SCM_h block can also be updated if the number of bits in the computer arithmetic is changed.

3.4.4 Block and Verilog Description of the 3D Lorenz System

The block diagram shown in Fig. 3.21 shows the description of the discretized equations given in Eq. (3.1) for Lorenz system. The use of the integrator block, adders, subtractors, multipliers, and SCMs can be appreciated.

One can see that in the block description shown in Fig. 3.21, the outputs are the three state variables at iteration x_{n+1}, y_{n+1}, and z_{n+1}. These values depend on one past value, i.e., at iteration x_n, y_n, and z_n, so that the forward Euler method is also known as a one-step and explicit method. In this block diagram, five clock cycles (CCs) are required to produce a valid data at the output. The details of the equation elements in each state variable are also given in the figure, and the Integrator_FE block is labeled for each state variable and SCM.

The Verilog description of the block representation shown in Fig. 3.21 is given in Listing 3.22. As one can see, the fixed-point format uses 32 bits distributed as listed

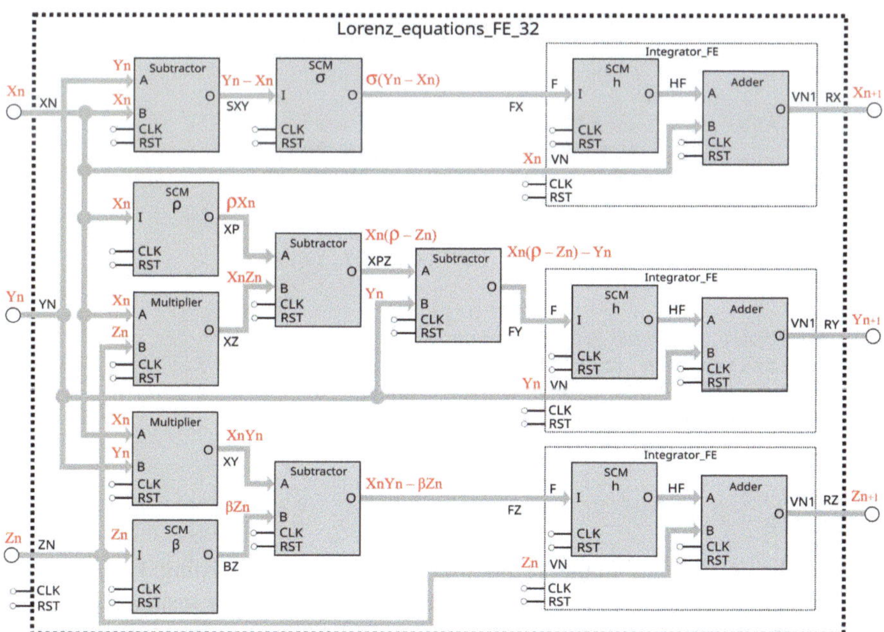

Fig. 3.21 Block description of the discretized equations given in Eq. (3.1) for the Lorenz system

in Table 3.1. Each equation of the three state variables is described by instantiation of multipliers, subtractors, SCMs, and the block Integrator_FE.

Listing 3.22 Verilog description of Lorenz system from Eq. (3.1) and from the hardware shown in Fig. 3.21

```
/*------------------------------------------------------------
---Project: Lorenz equations using Forward-Euler          ---
---Fixed-point format: 1:11:20                            ---
---Inputs: RST - sets the outputs to zero                 ---
           CLK - clock signal                             ---
           XN,YN,ZN - logic 32-bit inputs                 ---
---Output: RX,RY,RZ - logic 32-bit outputs                ---
------------------------------------------------------------*/
module Lorenz_equations_FE_32(RST,CLK,XN,YN,ZN,RX,RY,RZ);
    localparam n=32;
    localparam frac=20;

    input RST,CLK;
    input [n-1:0] XN,YN,ZN;
    output [n-1:0] RX,RY,RZ;

    wire [n-1:0] SYX,FX;
    wire [n-1:0] XP,XZ,XPZ,FY;
    wire [n-1:0] XY,BZ,FZ;

    /*---Description of equation x[n+1]---*/
         Subtractor #(n) U01 (RST,CLK,YN,XN,SYX);
                  SCM_10 U02 (RST,CLK,SYX,FX);
           Integrator_FE U03 (RST,CLK,FX,XN,RX);
    /*-------------------------------------------*/

    /*---Description of equation y[n+1]---*/
                  SCM_28 U04 (RST,CLK,XN,XP);
    Multiplier #(n,frac) U05 (RST,CLK,XN,ZN,XZ);
         Subtractor #(n) U06 (RST,CLK,XP,XZ,XPZ);
         Subtractor #(n) U07 (RST,CLK,XPZ,YN,FY);
           Integrator_FE U08 (RST,CLK,FY,YN,RY);
    /*-------------------------------------------*/

    /*---Description of equation z[n+1]---*/
    Multiplier #(n,frac) U09 (RST,CLK,XN,YN,XY);
                 SCM_8_3 U10 (RST,CLK,ZN,BZ);
         Subtractor #(n) U11 (RST,CLK,XY,BZ,FZ);
           Integrator_FE U12 (RST,CLK,FZ,ZN,RZ);
    /*-------------------------------------------*/

endmodule
```

It can be noticed that in the Verilog descriptions given in Listing 3.22, the SCM blocks, and Integrator_FE do not have input parameters in their instantiations. This is due to the fact that these modules are designed according to a computer arithmetic of fixed-point format, previously defined. If the descriptions are performed with a different number of bits or format, then the SCMs must be newly designed.

At this synthesis stage, the descriptions of the system of discretized equations do not have the feedback loop. In this manner, to perform the iterative process, the outputs are connected to the inputs to follow the next iterations [n+1]. This process is performed by adding additional blocks, some of them to store the initial conditions of the chaotic system. This process is detailed in Chaps. 5 and 6, for the design of PRNGs using integer- and fractional-order chaotic systems, respectively.

3.4.5 Block and Verilog Descriptions of a 2D Chaotic Map

The second case study is the 2D chaotic map introduced by Sprott [12]. This 2D map has two state variables, whose equations can be synthesized by using adders and multipliers. The discrete equations are already in the form of difference equations and are given in Eq. (3.2).

$$\begin{aligned} x_{n+1} &= a_1 + a_2 x_n + a_3 x_n^2 + a_4 x_n y_n + a_5 y_n + a_6 y_n^2 \\ y_{n+1} &= a_7 + a_8 x_n + a_9 x_n^2 + a_{10} x_n y_n + a_{11} y_n + a_{12} y_n^2 \end{aligned} \tag{3.2}$$

This chaotic map has 12 coefficients a_i, whose values can be changed to generate different chaotic attractors. Some works have already computed different sets of values, as the work given in [13], which provides four sets of values for the 12 coefficients, as listed in Table 3.2.

The following subsections show the block and Verilog descriptions of the 2D chaotic map, using the coefficient values of the second row in Table 3.2.

3.4.6 Selection of the Fixed-Point Format

Similar to the Lorenz system, before the development of the Verilog descriptions for the 2D chaotic map, one must simulate the equations to identify the maximum and minimum values of the state variables and arithmetic operations in the whole system of equations. In this manner, after simulating the 2D chaotic map with the

Table 3.2 Sets of values of the 12 coefficients a_i computed in [13], for the 2D Sprott map given in Eq. (3.2)

	Coefficient values											
Map	a_1	a_2	a_3	a_4	a_5	a_6	a_7	a_8	a_9	a_{10}	a_{11}	a_{12}
Map1	−0.6	−0.1	1.1	0.2	−0.8	0.6	−0.7	0.7	0.7	0.3	0.6	0.9
Map2	−1.0	0.9	0.4	−0.2	−0.6	−0.5	0.4	0.7	0.3	−0.5	0.7	−0.8
Map3	0.8	1.0	−1.2	−1.0	1.1	−0.9	0.4	−0.4	−0.6	−0.2	−0.5	−0.7
Map4	−0.6	−0.4	−0.4	−0.8	0.7	0.3	−0.4	0.4	0.5	0.5	0.8	−0.1

Table 3.3 Distribution of 64 bits in fixed-point format for the 2D Sprott map

Sign	Integer part	Decimal part
1 bit	4 bits	59 bits

set of values in the second row in Table 3.2, by performing 1,000,000 iterations with different initial conditions, the amplitudes are in the range ± 16. In the work [13], the 2D chaotic map is implemented in computer arithmetic of fixed-point format using 64 bits. In this book, the FPGA implementation is also performed by using 64 bits to design the blocks and their Verilog descriptions. The 64 bits are distributed as listed in Table 3.3, which considers the maximum and minimum amplitudes of ± 16.

3.4.7 Block and Verilog Descriptions of the Control of the Equations of the 2D Sprott Map

Chaotic maps are directly modeled by difference equations that can be directly synthesized by digital blocks. Analyzing the difference equations given in Eqs. (3.2), one can see two state variables with similar structure, but with different coefficient values a_i. The block description can be performed similar to the one performed for Lorenz system, by connecting adders, multipliers, and SCMs. However, since for this case study the descriptions are performed by using 64 bits, the number of embedded multipliers that are available into an FPGA could be high, thus causing the need of using a high number of hardware resources. One alternative can be the organization of the equations to use blocks like the one described in Sect. 3.3.10, which details the description of the multiplier accumulator (MAC) block.

To synthesize a set of difference equations by using MAC blocks, the first step is the definition of a finite-state machine (FSM) to control the number of products to be executed. In this case, the FSM must control the execution of six multiplication operations and cumulative sum. Figure 3.22 shows the internal hardware description for the control unit of the MAC and its black box representation. It can be appreciated that the MAC_control block consists of a Moore-type finite-state machine FSM_MAC and a counter Counter_opc, which has been detailed above.

The FSM_MAC embedded in Fig. 3.22 has four inputs and four control outputs. The outputs perform the following operations:

- OPC (operational counter): This bus controls two possible behaviors of the counter. If OPC =0, the counter is cleaned or reset to zero, and if OPC =2, the counter behaves in ascending fashion.
- LAC (load accumulator): Enables the multiplication block and the accumulator register.
- LFR (load final register): Enables the register that stores the final accumulated value.

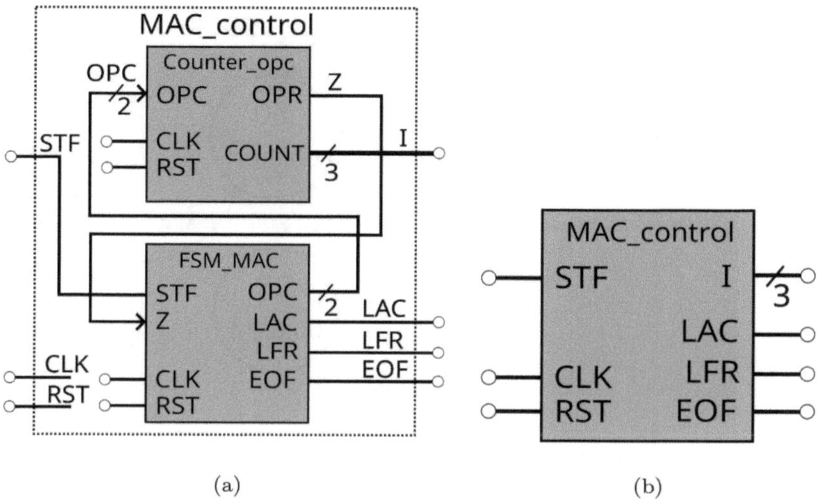

Fig. 3.22 Description of the block that controls the MAC: (**a**) internal hardware description and (**b**) black box representation

- EOF (End of function): Provides a logic 1 when the MAC block is ready to work.

On the other hand, the four inputs of the FSM_MAC block perform the following operations:

- STF (start function): starts the state machine
- Z (zero): input that receives a logic 1 when the multiplications previously defined have been accomplished or finished
- CLK (clock): pin that is assigned to the clock
- RST (reset): pin assigned to reset the block

The other block embedded into the MAC_control, the counter_opc provides the output I, which transfers the value of the current multiplication. The transitions of these states are sketched in Fig. 3.23, for the FSM_MAC block.

The FSM_MAC starts at state labeled S0. When signal STF is in a high logic state 1 and a clock cycle (CC) occurs, the FSM_MAC goes to state S1. At this state, the machine sets the outputs to OPC = 2 and LAC = 1 in order that the counter starts increasing to perform the multiplications and accumulation sums. To exit from state S1 and go to state S2, the counter_opc must send the signal Z, which occurs when all necessary multiplications are executed. At state S2 the output LFR is activated to get the final values that are accumulated for each state variable. Afterwards, the FSM_MAC returns to state S0 to wait for a new activation signal in STF.

With this control block for the MAC, it is only necessary to establish or set the counter to the number of multiplications to be performed. The Verilog description of the FSM_MAC is given in Listing 3.23, and Listing 3.24 shows the Verilog description for the block diagram shown in Fig. 3.22.

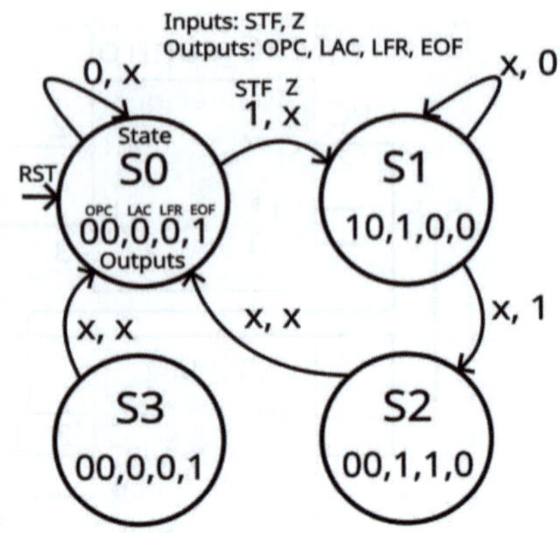

Fig. 3.23 Diagram of the transitions of the FSM_MAC block

Listing 3.23 Verilog description of the FSM_MAC block

```
 /*-----------------------------------------------------------------
---Project: Moore FSM_MAC                                        ---
---Inputs: RST - sets the initial state                          ---
           CLK - clock signal                                    ---
           STF,Z - logic inputs                                  ---
---Output: OPC,LAC,LFR,EOF - logic outputs                       ---
-----------------------------------------------------------------*/
module FSM_MAC(RST,CLK,STF,Z,OPC,LAC,LFR,EOF);
    localparam S0=0, S1=1, S2=2, S3=3;

    input RST,CLK,STF,Z;
    output reg [1:0] OPC;
    output reg LAC,LFR,EOF;
    reg [1:0] EDO;

    always @ (EDO) begin
        case (EDO)
            S0:
              begin OPC<=2'b00; LAC<=1'b0; LFR<=1'b0; EOF<=1'b1;
                  end
            S1:
              begin OPC<=2'b10; LAC<=1'b1; LFR<=1'b0; EOF<=1'b0;
                  end
            S2:
              begin OPC<=2'b00; LAC<=1'b1; LFR<=1'b1; EOF<=1'b0;
                  end
            S3:
              begin OPC<=2'b00; LAC<=1'b0; LFR<=1'b0; EOF<=1'b1;
                  end
            default:
              begin OPC<=2'b00; LAC<=1'b0; LFR<=1'b0; EOF<=1'b1;
                  end
        endcase
    end

    always @ (posedge CLK or posedge RST) begin
```

```
        if (RST) EDO <= S0;
        else
            case (EDO)
                S0:
                  if(STF)
                    EDO<=S1;
                  else
                    EDO<=S0;
                S1:
                  if(Z)
                    EDO<=S2;
                  else
                    EDO<=S1;
                S2:
                    EDO<=S0;
                S3:
                    EDO<=S0;
            endcase
    end
endmodule
```

Listing 3.24 Verilog description of the MAC_control block

```
/*--------------------------------------------------------------
---Project: Control for MAC block                            ---
---Inputs: RST - sets the initial state                      ---
           CLK - clock signal                                ---
           STF - start the function                          ---
           LAC - load accumulator register                   ---
           LFR - load final result                           ---
---Output: EOF - end of function                             ---
             I - current multiplication                      ---
--------------------------------------------------------------*/
module MAC_control #(parameter nc=3, parameter num_p=6)
    (RST,CLK,STF,LAC,LFR,EOF,I);
    input RST,CLK,STF;
    output LAC,LFR,EOF;
    output [2:0]  I;

    wire Z;
    wire [1:0] OPC;

                        FSM_MAC U01 (RST,CLK,STF,Z,OPC,LAC,LFR,EOF);
    Counter_opc #(nc,num_p) U02 (RST,CLK,OPC,I,Z);

endmodule
```

The creation of architectures by using MAC blocks embeds ROMs to store the constants that are used in the multiplication processes. The design of the MAC block that is used in the synthesis of the 2D Sprott map uses two ROMs. In this case memory ROM_X stores the constants from a_1 to a_6, of the first state variable, while memory ROM_Y stores the constants from a_7 to a_{12}. Both ROM_X and ROM_Y memories have the Verilog description given in Listing 3.14, but with different *.txt files.

Table 3.4 Constants a_1 to a_6 of the *.txt file for memory ROM_X, in fixed-point format 1:4:59

Address	Datum
0	1111100000000000000000000000000000000000000000000000000000000000
1	0000011100110011001100110011001100110011001100110011001101000000
2	0000001100110011001100110011001100110011001100110011001101000000
3	1111111001100110011001100110011001100110011001100110100000000000
4	1111101100110011001100110011001100110011001100110011000000000000
5	1111110000000000000000000000000000000000000000000000000000000000

Table 3.5 Constants a_7 to a_{12} of the *.txt file for memory ROM_Y, in fixed-point format 1:4:59

Address	Datum
0	0000001100110011001100110011001100110011001100110011001101000000
1	0000010110011001100110011001100110011001100110011001100110000000
2	0000001001100110011001100110011001100110011001100110011001100000
3	1111100000000000000000000000000000000000000000000000000000000000
4	0000010110011001100110011001100110011001100110011001100110000000
5	1111100110011001100110011001100110011001100110011001100000000000

The *.txt files for ROM_X and ROM_Y memories must store the values of the constants a_i in fixed-point format by using 64 bits. Each row in the file represents a memory cell and the number of columns represent the length of the digital word or datum. Considering this issue, two *.txt files must be generated with at least 6 rows and 64 columns. Table 3.4 shows the 6 rows with the binary values (in fixed-point format) for the file AX2.txt, and Table 3.5 shows the binary values for the file AY2.txt. These values of the constants a_i correspond to the ones for the 2D Sprott map in the second row (Map2) listed in Table 3.2.

As ROMs have 32 cells, the remaining rows in the *.txt files are filled with 64-bit words inserting zeros.

3.4.8 *Block and Verilog Description of the 2D Sprott Map*

The block diagram description shown in Fig. 3.24 shows the connection of the MAC block with the control one, the connection with ROM memories, a multiplexer8_1, and a multiplier-multiplexed (see Fig. 3.17); all the connections are performed to synthesize the two difference equations of the 2D chaotic map in a 64-bit architecture with fixed-point format.

As one can see, the block description shown in Fig. 3.24 provides the outputs x_{n+1} and y_{n+1}, which depend on the past values of the state variables, just at a previous iteration, labeled as x_n and y_n. To process one iteration from [n] to [n+1], nine clock cycles (CCs) are required. The Verilog description of the internal hardware shown in Fig. 3.24 is given in Listing 3.25.

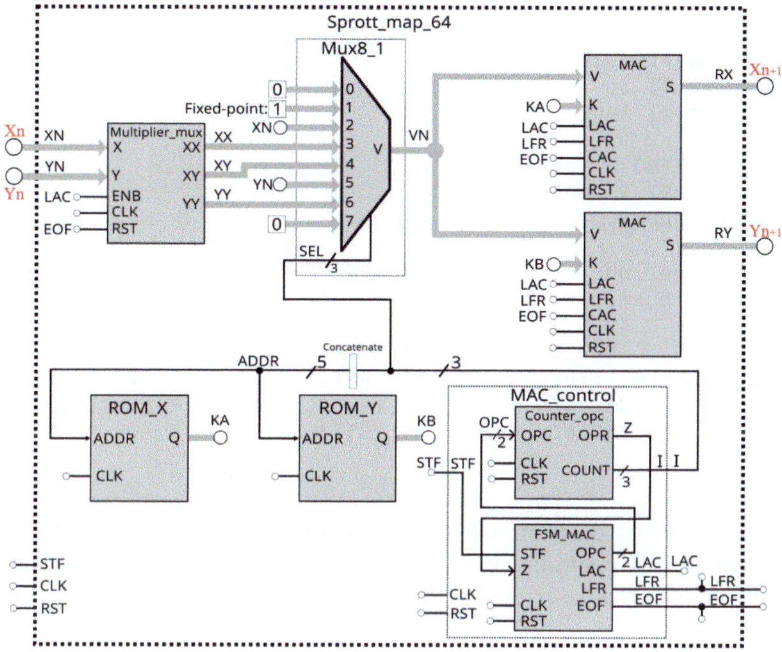

Fig. 3.24 Block description of the 2D Sprott map

Listing 3.25 Verilog description of the 2D Sprott_map

```
/*-----------------------------------------------------------------
---Project: Sprott map                                        ---
---Fixed-point format: 1:4:59                                  ---
---Inputs: RST - sets the outputs to zero                      ---
           CLK - clock signal                                  ---
           STF - start function                                ---
           XN,YN - logic 64-bit inputs                         ---
---Output: RX,RY - logic 64-bit outputs                        ---
           LFR - load final register                           ---
           EOF - end of function                               ---
-----------------------------------------------------------------*/
module Sprott_map_64(RST,CLK,STF,XN,YN,RX,RY,LFR,EOF);
    localparam n=64;
    localparam frac=59;
    localparam nc=3;
    localparam num_p=6;
    localparam onefix = 64'h0800000000000000; //One Fixed-point

    input RST,CLK,STF;
    input [n-1:0] XN,YN;
    output [n-1:0] RX,RY;
    output LFR,EOF;

```

```
    wire LAC;
    wire [n-1:0] UNI,ZERO,XX,XY,YY,VN,KA,KB;
    wire [n-1:0] RX,RY;
    wire [2:0] I;
    wire [4:0] ADDR;

    assign ZERO = {n{1'b0}};
    assign UNI =  onefix;
    assign ADDR = {2'b0,I};

    Multiplier_mux #(n,frac) U01 (EOF,CLK,LAC,XN,YN,XX,XY,YY);
              Mux8_1 #(n) U02 (I,ZERO,UNI,XN,XX,XY,YN,YY,ZERO,VN);
                 ROM_X #(n,5) U03 (CLK,ADDR,KA);
                 ROM_Y #(n,5) U04 (CLK,ADDR,KB);
     MAC_control #(nc,num_p) U05 (RST,CLK,STF,LAC,LFR,EOF,I);
                MAC #(n,frac) U06 (RST,CLK,LAC,LFR,EOF,VN,KA,RX);
                MAC #(n,frac) U07 (RST,CLK,LAC,LFR,EOF,VN,KB,RY);

endmodule
```

As for the Verilog description of the Lorenz system, the feedback loop of the 2D Sprott map must be controlled by adding control blocks, as detailed in Chap. 5 for the design of a PRNG based on this chaotic map.

3.5 Conclusion

This chapter showed the Verilog description of arithmetic and sequential blocks. The importance of describing combinational and arithmetic blocks such as adders, subtractors, and multipliers, using PIPO registers to have a sequential property was detailed. The addition of a clock pin to a combinational block is useful to synthesize equations that require the control of evaluations, as for the chaotic systems described in this book. Some examples of instantiation of basic blocks into bigger ones are given, as well as the design of the control circuitry, as the description of finite-state machines. The majority of blocks are used in the Verilog description of a 3D Lorenz system and a 2D chaotic map, in a higher level of design. These descriptions are instantiated in the design of PRNGs, as shown in Chaps. 5 and 6.

References

1. K.T. Chau, Z. Wang, *Chaos in Electric Drive Systems: Analysis, Control and Application* (Wiley, New York, 2011)
2. E. Tlelo-Cuautle, A.D. Pano-Azucena, O. Guillén-Fernández, A. Silva-Juárez, *Analog/Digital Implementation of Fractional Order Chaotic Circuits and Applications* (Springer, Cham, Switzerland, 2020)

3. Z. Kotadai, C. Fischer, J. Rodríguez-Muñoz, E. Tlelo-Cuautle, E. Yves, Multistability and initial-offset boosting dynamics in a new 3d memristive chaotic system with fpga implementation. Eur. Phys. J. Plus **139**, 70 (2024)
4. D. Thomas, P. Moorby, *The Verilog® Hardware Description Language* (Springer Science & Business Media, New York, 2008)
5. S.S. Chonnad, N.B. Balachander, *Verilog. Frequently Asked Questions: Language, Applications and Extensions* (Springer Science + Business Media, New York, 2004)
6. J. Rodríguez-Muñoz, J. Tavizón-Aldama, E. Tlelo-Cuautle, Experimental observation of chaotic attractors from the fpga implementation of 3d chaotic systems. IETE J. Educ., 1–9 (2023)
7. R.B. Reese, M.A. Thornton, *Introduction to Logic Synthesis Using Verilog HDL* (Springer Nature, New York, 2022)
8. U. Meyer-Baese, U. Meyer-Baese, *Digital Signal Processing with Field Programmable Gate Arrays*, vol. 65 (Springer, New York, 2007)
9. F. Vahid, *Digital Design with RTL Design, VHDL, and Verilog* (Wiley, New York, 2010)
10. B. Cohen, *Real Chip Design and Verification Using Verilog and VHDL* (vhdlcohen Publishing, Los Angeles, 2002)
11. P.D. Minns, I. Elliott, *FSM-Based Digital Design Using Verilog HDL* (Wiley, New York, 2008)
12. J.C. Sprott, Automatic generation of strange attractors. Comput. Graph. **17**(3), 325–332 (1993)
13. L.G. De la Fraga, C. Mancillas-López, E. Tlelo-Cuautle, Designing an authenticated hash function with a 2d chaotic map. Nonlinear Dynam. **104**, 4569–4580 (2021)

Chapter 4
Statistical Tests for PRNGs

4.1 NIST Randomness Test Suite

The National Institute of Standards and Technology (NIST) of the United States of America had created a free software for testing pseudo- and random number generators [(P)RNGs]. Specifically, NIST said that the test is for (P)RNGs for cryptographic applications, where randomness is a crucial characteristic. The NIST suite package is available at,[1] and according to NIST [1, Ch. 5] it can address the problem of evaluating (P)RNGs for randomness. The usefulness of this statistical test suite is that it was created to help in identifying (P)RNGs which produce weak (or patterned) binary sequences, designing new (P)RNGs, verifying that the implementations of (P)RNGs are correct, studying (P)RNGs described in standards, and investigating the degree of randomness of the currently used (P)RNGs.

4.1.1 How to Use NIST Test Suite

The generation of binary strings needs to be evaluated in order to verify the degree of randomness. One way is the application of NIST tests to sequences generated by chaotic maps [2–4] or by continuous time chaotic oscillators [5, 6], as the case study in this book.

In performing NIST tests, it is common to use a set of 100 sequences, each one of 10^6 bit length. This chapter shows the application of the suite to some sequences. For instance, the NIST test source code is freely available in C language. Once it is compiled, using gcc, its execution can be performed as follows:

```
./assess 1000000,
```

[1] http://csrc.nist.gov/groups/ST/toolkit/rng/documentation_software.html

L. G. de la Fraga et al., *Random Number Generators*,
https://doi.org/10.1007/978-3-031-82865-2_4

where `assess` is the program that applies the statistical tests, and `1000000` is the binary sequence length to be tested. In this book, the tested sequences are equal to one million (10^6) bits in length. It is quite important to provide the sequence length with an integer number. Once the showed command is executed, the following options are printed in the terminal window:

```
$ ./assess 1000000
        G E N E R A T O R    S E L E C T I O N
        ______________________________________

  [0] Input File                  [1] Linear Congruential
  [2] Quadratic Congruential I    [3] Quadratic Congruential II
  [4] Cubic Congruential          [5] XOR
  [6] Modular Exponentiation      [7] Blum-Blum-Shub
  [8] Micali-Schnorr              [9] G Using SHA-1

 Enter Choice:
```

It is necessary to key “0” and after that key “enter” to select the input file. Then, the following text is printed to the terminal:

```
    P a r a m e t e r   A d j u s t m e n t s
     -----------------------------------------
  [1] Block Frequency Test - block length(M):         128
  [2] NonOverlapping Template Test - block length(m): 9
  [3] Overlapping Template Test - block length(m):    9
  [4] Approximate Entropy Test - block length(m):     10
  [5] Serial Test - block length(m):                  16
  [6] Linear Complexity Test - block length(M):       500
```

where six tests and their corresponding six parameter values in the right are shown. If these values are correct, they must be checked in the NIST manual considering the input sequence size and number of sequences (which will be given as the last parameter before running the `assess` command.

In this section, an example is described to perform NIST statistical tests by considering 100 binary sequences, each one having a length of 1 million bits, to set the input value parameters. The first NIST test, called `Block Frequency Test`, calculates statistics with a fixed block of size; in the documentation [1] it is suggested that $M \geq 20$, $M \geq 0.01n$, and $N < 100$, where n is the input size sequence and $n \geq MN$, and then with $M = 16{,}384 = 2^{14}$, one can have $M \geq 20$, $M \geq 0.01(10^6) = 10^4 < 16{,}384$, $N = n/M = 10^6/16{,}384 = 61.035$ which is lower than 100.

For the second NIST test, `NonOverlapping Template Test`, the suggested values for parameter m can be established as 9 or 10.

For the third NIST test, `Overlapping Template Test`, the NIST statistical test suite recommends that the input sequences must have a minimum length of 10^6 bits, and m should be equal to 9 or 10.

For the fourth NIST test, `Approximate Entropy Test`, one must choose m and n such that $m < \lfloor \log_2 n \rfloor - 5$ is accomplished; for example, $\lfloor \log_2 10^6 \rfloor = 19$ and $m < 19 - 5 = 14$. This is the maximum value that m can take. For the experiments performed in this book for the binary sequences generated by chaotic maps and integer- and fractional-order chaotic systems, the value of m is chosen to be equal to 12.

For the fifth NIST test, `Serial Test`, the required parameters m and n must be chosen to accomplish that $m < \lfloor \log_2 n \rfloor - 2$. In the example provided in this book, the values that are used in the length of bits produce $\lfloor \log_2 10^6 \rfloor = 19$; thus, $m < 19 - 2 = 17$.

Finally, for the sixth NIST test, `Linear Complexity Test`, the documentation recommends that n must be chosen such that $n \geq 10^6$, and then the value of M must be within the range $500 \leq M \leq 5000$ and $N \geq 200$. This produces that one can have $N = n/M$, for $n = 10^6$, $N = 10^6/500 = 2000$. If $M = 1000$, one can have $N = 10^6/1000 = 1000$. Then, for a binary sequence length of $n = 10^6$ bits, the parameters must be chosen as it is indicated in Table 4.1.

To avoid the manual interaction with the `assess` program, it is possible to use it automatically within a loop, the only requirement is that all the parameter values must be provided within a file. The command line execution is as follows:

```
./assess 1000000 < input.in
```

where file `input.in` has the content shown in Fig. 4.1, `1000000` is the number of bits per sequence (one million in this case), and the binary file with the generated zero and one bits is in the location `data/a.bin`. In this way, it is possible to automatically use the `assess` program, i.e., in a noninteractive way.

Let us consider the 2D Sprott map already used in the work [4]; from this map one can generate 100 sequences of 1 million bits in length (the file has a length of 12,500,000 bytes). The execution of the NIST test suite produces two results of `CumulativeSums`, 148 results for `NonOverlappingTemplate` test, 8 results for `RandomExcursions` test, 18 results for `RandomExcursionsVariant` test, and 2 results for `Serial` test. Reading research works that include NIST tests, it is common to observe that one can find average values of each all those tests. In particular, tests `RandomExcursions` and `RandomExcursionsVariant` are applied on 69 sequences, and also, in the literature it is common to observe that authors

Table 4.1 Values for the NIST statistical tests that are performed in this book.

Test	Parameter value
1	$M = 2^{14} = 16{,}384$
2	$m = 9$ or $m = 10$
3	$m = 9$ or $m = 10$
4	$m = 12$
5	$m = 17$
6	$M = 500$

```
0            % Enter input file
data/a.bin   % Location and name of the input file
1            % Apply all tests
1            % 1st test,
16384        % parameter value for the 1st test.
4            % 4th test, "Approximate Entropy Test",
12           % set block lenght iqual to 12.
5            % 5th test, "Serial Test",
17           % block lenght equal to 17.
0            % To continue
100          % The number of sequences
1            % Selects binary input file
```

Fig. 4.1 Contents of `input.in` file to avoid using the program `assess` in an interactive fashion Text to the right starting with the symbol `%` are comments. Be aware that file `input.in` must not contain those comments

Table 4.2 Results of applying the 15 NIST statistical tests on 1000 sequences of 10^6 bits to the PRGN proposed in [4]. A test passes if the proportion value is greater than 96 and the p-value is greater than 0.0001 [1, Sec. 4.2.2]

	Test name	p-Value	Proportion
1	Frequency	0.042808	99
2	BlockFrequency	0.739918	98
3	CumulativeSums	0.083665	99
4	Runs	0.834308	98
5	LongestRun	0.759756	99
6	Rank	0.595549	98
7	FFT	0.924076	99
8	NonOverlappingTemplate	0.476494	99
9	OverlappingTemplate	0.419021	98
10	Universal	0.657933	98
11	ApproximateEntropy	0.851383	97
12	RandomExcursions	0.196092	99
13	RandomExcursionsVariant	0.316751	99
14	LinearComplexity	0.350485	99
15	Serial	0.568117	99

show the average normalized results taking these 69 sequences as the 100% of the sequences.

The output file with the NIST test results is generated and located at the directory `experiments/AlgorithmTesting` in the file `finalAnalysisReport.txt`. The final report of the NIST test suite for this example is shown in Table 4.2. Some insights that must be considered are listed as follows: A test passes the proportion value if it is greater than 96 (for 100 sequences), and the p-value shown in Table 4.2 is greater or equal than 0.0001 [1, Sec. 4.2.2]. The p-value is related to how the p-values for the tests in all the 100 sequences are uniformly distributed. The p-value used with each test is equal to 0.01, and it is also important to check if the distribution of all those p-values are uniform and then the sequences can be considered random. Additionally, in the guidelines provided in [1], one can find that the provided statistically meaningful results should validate that at least 55 sequences must be processed.

4.2 TestU01 Test Suite

Another statistical test suite that is also recommended to be applied for the verification of randomness of chaotic binary strings, as the ones generated by the PRNGs designed in Chaps. 5 and 6, is the TestU01. Detailed information of this test suite can be found in the web page http://simul.iro.umontreal.ca/testu01/tu01.html. This TestU01 suite is not executed as a command, as for the case described for the NIST test suite. This test suite is a library that is programmed in C language. Basically, three tests of TestU01 suite are used to verify randomness of binary sequences, namely, Rabbit, Alphabit, and BlockAlphabit. These three tests were designed for strings of bits stored in a file (or a physical device). Rabbit and Alphabit apply 38 and 17 different statistical tests, respectively. BlockAlphabit applies the Alphabit battery of tests repeatedly to a binary file after reordering the bits by blocks of different sizes (with sizes of 2, 4, 8, 16, 32 bits) [7]. The source code to apply the Rabbit test is given as follows:

```
#include <stdio.h>
#include <stdlib.h>
#include "ulcg.h"
#include "unif01.h"
#include "bbattery.h"

int main( int argc, char * argv[] )
{
    double nb;

    if ( argc != 3 ) {
        fprintf( stderr, "Args: file_name nbits\n" );
        return 1;
    }
    nb = strtod( argv[2], NULL );
    bbattery_RabbitFile ( argv[1], nb );

    return 0;
}
```

Here if the created new command is called `bat1`, the two arguments that are passed to it are the binary file name and the number of bits within the file. These two arguments are passed directly to the function `bbattery_RabbitFile()` which does the Rabbit test. For the other two tests, Alphabit and BlockAlphabit, other two commands can be created by using the same source code as the one given above, but now by changing to the function names `bbattery_AlphabitFile()` and `bbattery_BlockAlphabitFile()`, respectively.

Similar to the NIST test suite, the application of these three TestU01 tests to 100 sequences of 10^6 bits can be performed as follows: All the bits in the binary string must be stored in a file named `a.bin`. Recall that the binary strings have been generated, in this chapter, by using the 2D Sprott map described in Chap. 3 [4], but one can use other binary strings, as the ones generated by the integer- and fractional-order chaotic systems given in Chaps. 5 and 6. Each suite for the TestU01 tests is applied by doing the corresponding assignation of the binary string (`s?.txt`), by typing the following command line:

```
./bat1 a.bin 1e8 > s1.txt
./bat2 a.bin 1e8 > s2.txt
./bat3 a.bin 1e8 > s3.txt
```

One python script was created to process each `s?.txt` file to count all the passed and the non-passed tests. The final results of applying TestU01 tests to the binary strings generated by using the 2D Sprott map are shown in Table 4.3.

4.2.1 TestU01 Test Suite on Floating Point Numbers in [0, 1)

There exist batteries of TestU01 tests that can be applied on sequences of uniform floating-point numbers within the interval [0, 1). In this case, one can apply the four available C functions, which are labeled as follows:

```
void bbattery_SmallCrush(unif01_Gen gen);
void bbattery_SmallCrushFile (char *filename);
void bbattery_Crush (unif01_Gen *gen);
void bbattery_BigCrush (unif01_Gen *gen);
```

The batteries listed above are called `SmallCrush`, `Crush`, and `BigCrush`. According to the TestU01 manual [7], "one could first apply the small and fast battery SmallCrush, if it passes, one could then apply the more stringent battery Crush, and finally the yet more time-consuming battery BigCrush." The only TestU01 test that uses an input file is the one labeled as `SmallCrush`. An important thing to consider is that this file requires slightly less than 51,320,000 random numbers [7], and then the file $51 \times 10 \approx 510$ MB. This value is obtained by considering that each float number is represented with 10 ASCII characters; then it is better to use the functions that call internally the binary sequences generated from

Table 4.3 Results of applying three TestU01 on 100 sequences of 10^6 bits by using the 2D Sprott map described in Chap. 3 [4]

	Test name	
1	Rabbit	All 40 tests passed
2	Alphabit	All 17 tests passed
3	Block Alphabit	All 6 repetitions of Alphabit tests passed

a PRNG. The following C code is an example on how to apply the `SmallCrush` on the C random generator `rand()`.

```
#include <stdlib.h>
#include "unif01.h"
#include "bbattery.h"

double myGenerator( void )
{
    return( (double)rand()/RAND_MAX );
}

int main (void)
{
    unif01_Gen *gen;

    srand( 20 );
    gen = unif01_CreateExternGen01 ( "myGenerator", myGenerator )
        ;
    bbattery_SmallCrush (gen);
    unif01_DeleteExternGen01 (gen);

    return 0;
}
```

The execution of these TestU01 tests produced the following summary results:

```
========= Summary results of SmallCrush =========

 Version:          TestU01 1.2.3
 Generator:        myGenerator
 Number of statistics:  15
 Total CPU time:   00:00:05.88
 The following tests gave p-values outside [0.001, 0.9990]:
 (eps  means a value < 1.0e-300):
 (eps1 means a value < 1.0e-15):

       Test                          p-value
 ----------------------------------------------
  3  Gap                              7.0e-8
  7  WeightDistrib                     eps
 ----------------------------------------------
 All other tests were passed
```

These summary results of SmallCrush indicate that two statistical tests of 15 failed. Thus, one can see that the built-in PRNG in the C language is not a good PRNG; indeed the main page of rand function suggests to use instead the random function in "applications intended to be portable when good randomness is needed."

The execution time of this TestU01 test was computed as 6 seconds in a laptop with the following characteristics: Intel Core-i5 (8 cores) CPU at 1 GHz, 8 GB RAM, with Debian 10.13 operating system.

To test a PRNG or RNG that produces bits, the function that must be used can be typed as follows:

```
unif01_Gen *unif01_CreateExternGenBits (char *name,
                              unsigned int (*genB)(void));
void unif01_DeleteExternGenBits (unif01_Gen * gen);
```

The function pointed by `genB` must return 32 random bits in the returned `unsigned int` variable.

Applying the **Crush** test to the same PRNG available in C produces the following summary results:

```
========= Summary results of Crush =========

 Version:          TestU01 1.2.3
 Generator:        myGenerator
 Number of statistics:  144
 Total CPU time:   00:26:27.24
 The following tests gave p-values outside [0.001, 0.9990]:
 (eps  means a value < 1.0e-300):
 (eps1 means a value < 1.0e-15):

       Test                          p-value
 ----------------------------------------------
  9  CollisionOver, t = 20          1 -  6.7e-7
 25  SimpPoker, d = 64                eps
 26  SimpPoker, d = 64                eps
 29  CouponCollector, d = 16        1.2e-10
 30  CouponCollector, d = 16         6.7e-7
 31  Gap, r = 0                       eps
 32  Gap, r = 27                      eps
 33  Gap, r = 0                       eps
 34  Gap, r = 22                      eps
 51  WeightDistrib, r = 0             eps
 52  WeightDistrib, r = 8             eps
 53  WeightDistrib, r = 16            eps
 54  WeightDistrib, r = 24            eps
 72  LinearComp, r = 29             1 - eps1
 ----------------------------------------------
 All other tests were passed
```

Analyzing the summary results of Crush given above, one can observe that these results say that 144 tests were applied and from them 14 failed. The **Crush** test took around 26.5 minutes on the same hardware as before.

4.3 Dieharder Test Suite

Another statistical test suite, different from NIST and TestU01, is the one called Dieharder. In fact, the original Diehard tests are still available in the following link: https://ani.stat.fsu.edu/diehard/ [8]. This statistical test is also available in GNU/Linux repositories with the name of *Dieharder*. The installation of Dieharder test suite can be performed by doing the following:

```
$ apt-cache search dieharder
dieharder - Random-number generator test front-end
libdieharder-dev - Random-number generator test library
-- development package
libdieharder3 - Random-number generator test library
$ sudo apt install dieharder
```

Diehard contains a large number of statistical tests. However, it has some drawbacks and limitations. Firstly, the sequence of tests to be applied to any generator, as well as the parameters of these tests (sample size, etc .), is fixed in the package. The sample sizes are moderate; all these tests run in a few seconds of CPU time on a desktop computer. Secondly, the package requires that the random numbers to be tested are 32-bit integers, placed in a huge file in binary format. This file is passed to the testing procedures. This setup is not always convenient because it is well known that not only FPGA implementations of PRNGs based on chaotic maps or systems, but also other hardware or software RNGs produce numbers with a length of bits lower than 32 bits of resolution (e.g., 31 bits is frequent) and Diehard does not care for that. It can be observed that different from Diehard, the TestU01 test suite is more flexible on all these aspects [7].

Dieharder test suite is available in TestU01, and it can be called by typing the function:

```
void bbattery_pseudoDIEHARD (unif01_Gen *gen);
```

This instruction is typed to apply the battery PseudoDIEHARD, which implements most of the tests listed in the Diehard suite [8] or, in some cases, performs close approximations to them. TestU01 authors do not recommend this battery, as it has been observed that it is not very stringent (they do not know of any generator that passes the batteries Crush and BigCrush and fails PseudoDIEHARD, while one can see the conversion for several defective generators).

The application of Dieharder test suite requires a huge quantity of bytes. As an example, the output to apply this suite test to a file of 1 million bits is described as follows:

```
$ dieharder -a -g 201 -f a.bin
#=============================================================================#
#            dieharder version 3.31.1 Copyright 2003 Robert G. Brown          #
#=============================================================================#
   rng_name    |           filename             |rands/second|
 file_input_raw|                           a.bin|  5.06e+07  |
```

```
#=============================================================================#
        test_name   |ntup| tsamples |psamples|  p-value |Assessment
#=============================================================================#
# The file file_input_raw was rewound 4 times
   diehard_birthdays|   0|       100|     100|0.71461020|  PASSED
# The file file_input_raw was rewound 36 times
      diehard_operm5|   0|   1000000|     100|0.00961315|  PASSED
# The file file_input_raw was rewound 77 times
  diehard_rank_32x32|   0|     40000|     100|0.15584182|  PASSED
# The file file_input_raw was rewound 96 times
    diehard_rank_6x8|   0|    100000|     100|0.91421007|  PASSED
# The file file_input_raw was rewound 104 times
   diehard_bitstream|   0|   2097152|     100|0.10791960|  PASSED
```

It can be appreciated that in the text given above, one can see only the first six tests. The first test, `diehard_birthdays`, rewound the input file four times! And the next tests even use a lot of rewound on the input file.

4.4 Conclusion

In this chapter, one can see some insights on the application of the three commonly used statistical test suits to probe if a sequence of bits or floating point numbers generated by a pseudo-random or random number generator is random.

After observing the test results, one can infer that it is important to take into account that the statistical tests can only conclude if the test is passed or not. Thus, one can only say if a sequence of bits or floating point number are random or not. In other words, one can conclude that there does not exist a classification or grade of randomness.

Some useful insights from practical experiences of the authors are the following: It is better to apply first the TestU01 suite to binary strings, because they take less time to test sequences compared to NIST or Dieharder test suites. After that, the NIST suite can be applied. Another insight is that to test sequences of floating point numbers in the interval [0, 1), only the TestU01 suite is available.

References

1. A. Rukhin, J. Soto, J. Nechvatal, M. Smid, E. Barker, S. Leigh, M. Levenson, M. Vangel, D. Banks, A. Heckert, J. Dray, S. Vo, A statistical test suite for the validation of random number generators and pseudo random number generators for cryptographic applications (April 27, 2010). SP800-22rev1a.pdf file, last date checked: Jan 31, 2024. Online at http://csrc.nist.gov/groups/ST/toolkit/rng/documentation_software.html
2. L.G. de la Fraga, B. Ovilla-Martínez, A chaotic prng tested with the heuristic differential evolution. Integr. VLSI J. **90**, 22–26 (2023)
3. L.G. de la Fraga, B. Ovilla-Martínez, Generating even more chaotic instances in hardware. Electronics **12**, 332 (2023)

4. L.G. de la Fraga, C. Mancillas-López, E. Tlelo-Cuautle, Designing an authenticated hash function with a 2d chaotic map. Nonlinear Dynam. **104**, 4569–4580 (2021)
5. L.G. de la Fraga, B. Ovilla-Martínez, Generating pseudo-random numbers with a brownian system. Integration **96**, 102135 (2024)
6. O. Guillén-Fernández, E. Tlelo-Cuautle, L.G. de la Fraga, Y. Sandoval-Ibarra, J.-C. Nuñez-Perez, An image encryption scheme synchronizing optimized chaotic systems implemented on raspberry pis. Mathematics **10**(11), 1907 (2022)
7. L. Ecuyer, R. Simard, TestU01: A C library for empirical testing of random number generators. ACM Trans. Math. Software **33**(22), 1–40 (2007). http://simul.iro.umontreal.ca/testu01/tu01.html
8. G. Marsaglia, The marsaglia random number CDROM including the DIEHARD battery of tests of randomness (1996). See http://stat.fsu.edu/pub/diehard

Chapter 5
PRNGs Based on Chaotic Maps and 3D, 4D, and 5D Chaotic Systems

5.1 Introduction

Chaotic maps have proven their usefulness in designing pseudo-random number generators (PRNGs), for which their FPGA implementation requires low hardware resources, as shown in [1]. In that work, the authors generated binary pseudo-random sequences using three well-known 1D maps, namely, the Bernoulli shift, tent, and zigzag maps. The float values of the samples at each iteration were compared to a threshold to generate binary sequences. In addition, to increase randomness, the generated sequences were post-processed using the bit counting redundancy reduction technique [2], which takes the original sequence to be divided into blocks of 5 bits, and then a new bit for a new sequence is generated by applying the XOR operation on the 5 bits of each block. However, this post-processing procedure to generate binary random numbers is very inefficient because in five iterations only a single bit is generated. The PRNGs that are described in this chapter are created with the idea to produce more efficient ones and are based on chaotic binary sequences generated from chaotic maps and chaotic and hyperchaotic systems.

In this book, the procedure to generate PRNGs is inspired in fact on grayscale images, as the ones shown in Fig. 5.1. In these images, it can be observed that the least significant bits (LSBs) have a random pattern for the last bits 1 and 0. Similarly, John von Newman in 1946 suggested a method called *the middle-square method*, whose algorithm works as follows: Take any number, square it, remove the middle digits of the resulting number as the "random number," and then use that number as the seed for the next iteration. For example, squaring the number 1234 yields 1522756, which can be written as 01522756, an 8-digit number being the square of a 4-digit number. This gives 5227 as the "random" number. Repeating this procedure gives 3215 as the next result and so on. Von Neumann used 10-digit numbers, but the process was the same. This method has already proved to be a comparatively poor source of generating random numbers. The danger is that the sequence tends to get

L. G. de la Fraga et al., *Random Number Generators*,
https://doi.org/10.1007/978-3-031-82865-2_5

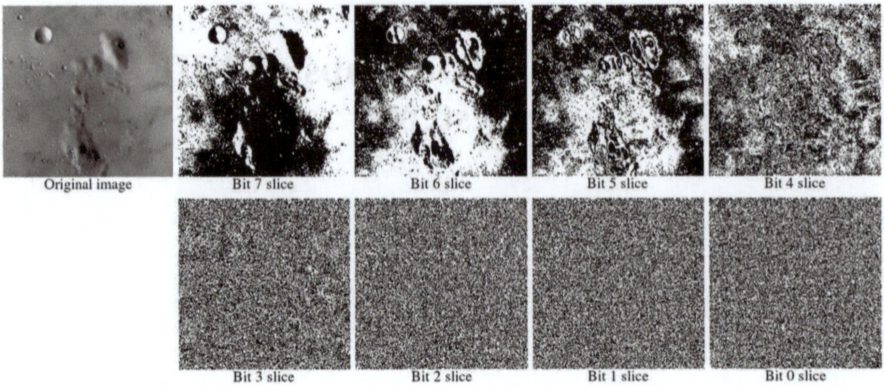

Fig. 5.1 A grayscale image and its eight bit-slice images. Each pixel of a grayscale image is represented by a byte

into a rut, a short cycle of repeating elements. For example, if zero ever appears as a number of the sequence, it will continually perpetuate itself [3, Ch. 3].

Therefore, to increase the randomness of PRNGs, the main idea in this book is to use large numbers, perform all operations in fixed-point arithmetic, and take the LSBs of the state variables that have buses of 32 or 64 bits, as the random bits. In this manner, the maps or chaotic systems are seen as an engine to introduce entropy at each iteration of the map or the numerical integration in a chaotic system.

All the PRNGs described in this chapter are based on either chaotic maps or chaotic systems and use numbers of type $a.b$, where a is taken to describe the number of bits for the integer part, and therefore, b is taken to describe the number of bits for the decimal part, so that $a+b=63$ bits. As the numbers are positive and negative, one bit is necessary to represent the sign, so that the computer arithmetic is performed by using 64 bits. Two single operations are needed here to implement the PRNGs and the sum and the multiplication of two numbers $a.b$. On the one hand, in a computer with a 64-bit architecture and using the C programming language, the type `long` is a 64-bit number. Thus, the sum can be performed simply summing up two `long` numbers. On the other hand, the multiplication of two numbers is a bit complicated: For example, the multiplication of two 64-bit numbers could result in a 128-bit number, and this result must be returned to an $a.b$ number. Using the gcc compiler, there exists the type `__int128` that can store a 128-bit integer. Then the multiplication code is as follows:

```
long multrunc( long n1, long n2 )
{
    __int128 r, num1, num2;

    num1 = n1;
    num2 = n2;
    r = num1*num2;
    return r>>b;
}
```

5.2 PRNG Based on a 2D Sprott Map

Chaotic maps can be described as 1D or 2D. In the case of having two state variables, a 2D map was proposed by Sprott in 1993 [4]. This 2D map is quite useful not only for generating random binary sequences but also for authenticating encrypted data, as shown in [5]. The difference equations of the 2D Sprott map are described by Eq. (5.1), which includes 12 coefficients labeled as a_i, and both variables are squared.

$$\begin{aligned} x_{n+1} &= a_1 + a_2 x_n + a_3 x_n^2 + a_4 x_n y_n + a_5 y_n + a_6 y_n^2, \\ y_{n+1} &= a_7 + a_8 x_n + a_9 x_n^2 + a_{10} x_n y_n + a_{11} y_n + a_{12} y_n^2. \end{aligned} \tag{5.1}$$

As already shown in [4], the coefficient values a_i can be searched randomly. However, chaotic behavior has been observed if the coefficients are varied within the range $[-1.2, 1.2]$, with an increment of 0.1. One set of values of the coefficients to generate a strange attractor, and which are used in this chapter, is the following:

$$\begin{aligned} \{ & -0.6, -0.4, -0.4, -0.8, 0.7, 0.3, \\ & -0.4, 0.4, 0.5, 0.5, 0.8, -0.1\}, \end{aligned}$$

With these a_i values, the 2D map has the Lyapunov exponent equal to 0.13 [4]. The domain of attraction of this 2D map is shown in Fig. 5.2.

As the FPGA implementation of PRNGs based on chaotic maps and systems is based on the fixed-point format, in this chapter the design of a PRNG with these

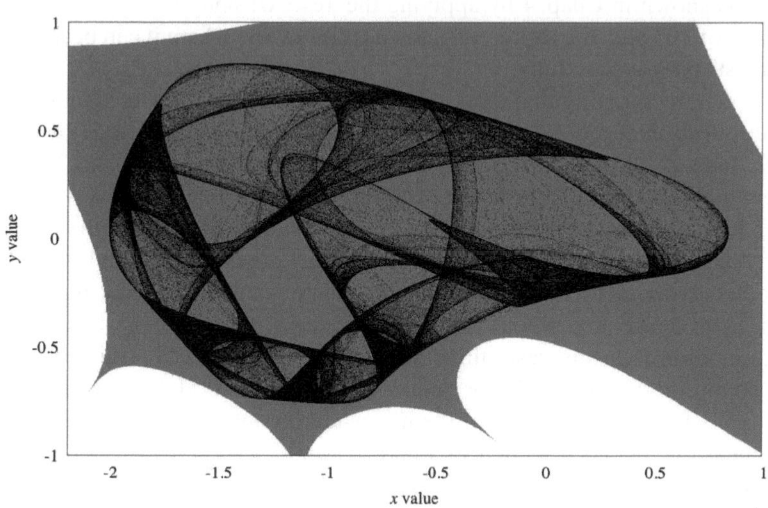

Fig. 5.2 The 2D chaotic map used in this chapter and its domains of attraction in floating-point arithmetic

a_i values requires to evaluate the maximum and minimum amplitudes to establish the number of bits that are used to develop the Verilog descriptions, as done in Chap. 3. It is then necessary to calculate first the number of bits that will be used in the integer part of the used numbers [6]. For instance, the maximum value for all the coefficients will be considered equal to 1, the maximum value for the variable x will be 2, and the maximum value for the variable y will be 1 (see the domain of attraction in Fig. 5.2). In this way, the maximum values for x_{n+1} and y_{n+1} are evaluated as follows:

$$x_{n+1} = 1 + 1(2) + 1(2)^2 + 1(2)(1) + 1(1) + 1(1)^2,$$

$$y_{n+1} = 1 + 1(2) + 1(2)^2 + 1(2)(1) + 1(1) + 1(1)^2,$$

$$x_{n+1} = 1 + 2 + 4 + 2 + 1 + 1 = 11,$$

$$y_{n+1} = 1 + 2 + 4 + 2 + 1 + 1 = 11.$$

The maximum amplitude values for both state variables are 11; therefore, it is necessary to use $2^4 = 16 > 11$, 4 bits to represent the integer part of the number in fixed-point format. As one bit is assigned to represent the sign, then the remaining bits to represent the decimal part are $63 - 4 = 59$ bits. Thus, the fixed-point format is described as 1:4:59.

The 12 a_i coefficients given in Eq. (5.1) are initialized using all the 59 bits for the decimal part. This initialization cannot be done by converting directly a real number to a 1:4:59 number because the floating-point numbers of double precision in C language have only 52 bits of precision.

The binary sequence can be generated by concatenating the 8 LSBs of the variables x_{n+1} and y_{n+1}. In this manner, the randomness of the binary sequences is tested as shown in Chap. 4 by applying the TestU01 and NIST test suites to 100 sequences of 10^6 bits; the results are shown in Table 5.1, where it can be appreciated that all tests pass successfully.

It is worth mentioning that there are 144 possible ways to associate the terms in the multiplication of the 2D map [7] to calculate the domains of attraction shown in Fig. 5.2. Indeed, all these 144 possible sets of values associated to the 2D map are different and therefore can generate different binary sequences. This phenomenon arises because the multiplication in the computer does not follow the associative property, and the roundoff effects are magnified by the chaotic behavior.

Another experiment is realized in this section to measure how the generated sequences are correlated when they are generated with the smallest change in the initial conditions. In this case, three sequences are generated with the following initial conditions: $[-1.5, 0.5]$ for sequence 1, $[-1.5\epsilon, 0.5]$ for sequence 2, and $[-1.5, 0.5\epsilon]$ for sequence 3, all of them considering an $\epsilon = 2^{-59}$. The correlation can be measured between sequences 1 and 2, 1 and 3, and between sequences 2 and 3, and the results can be shown as in Fig. 5.3. As one can observe, the sequences start to be uncorrelated around byte 90, or around bit 720; thus, this quantity of bits must be discharged when initial conditions are changed.

Table 5.1 Results of applying three TestU01 and fifteen NIST statistical tests to the 2D map to 100 sequences of 10^6 bits. The initial state was set equal to $[-1.5, -0.5]$

	Test name	p-value	Proportion
1	Rabbit	All 40 tests passed	
2	Alphabit	All 17 tests passed	
3	Block Alphabit	All 6 repetitions of Alphabit tests passed	
4	Frequency	0.759756	99
5	BlockFrequency	0.319084	100
6	CumulativeSums	0.914740	100
7	Runs	0.319084	100
8	LongestRun	0.798139	100
9	Rank	0.153763	99
10	FFT	0.096578	99
11	NonOverlappingTemplate	0.491413	99
12	OverlappingTemplate	0.366918	100
13	Universal	0.883171	99
14	ApproximateEntropy	0.657933	98
15	RandomExcursions	0.523318	99
16	RandomExcursionsVariant	0.660296	99
17	LinearComplexity	0.062821	100
18	Serial	0.495042	99

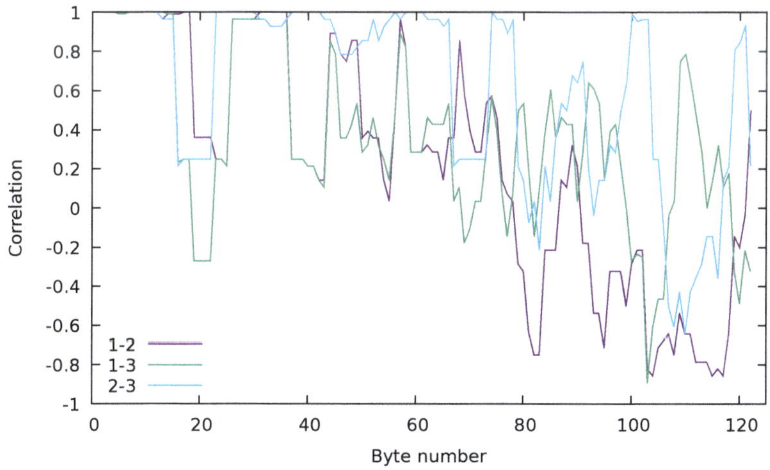

Fig. 5.3 Correlation among three sequences generated with different initial conditions: $[-1.5, 0.5]$ for sequence 1, $[-1.5 + \epsilon, 0.5]$ for sequence 2, and $[-1.5, 0.5 + \epsilon]$, for sequence 3, with $\epsilon = 2^{-59}$

5.3 FPGA Synthesis of a PRNG Based on 2D Sprott Map

The 2D Sprott map [4] has two state variables given in Eq. (5.1). In Chap. 3, one can see that Table 3.2 shows four sets of values of the 12 coefficients a_i that were computed in [5], for the 2D Sprott map. Figure 5.4 shows the chaotic time series for both variables x_n and y_n using the set of values a_i that correspond to Map1 (first row in Table 3.2) by using the initial conditions $x_0 = 0.1$ and $y_0 = 0.2$. The plots show the first 1,000 iterations of each state variable.

The chaotic attractors between both state variables for the four sets of values given in Table 3.2 are shown in Fig. 5.5. In these cases, all the phase portraits are plotted taking 10,000 iterations.

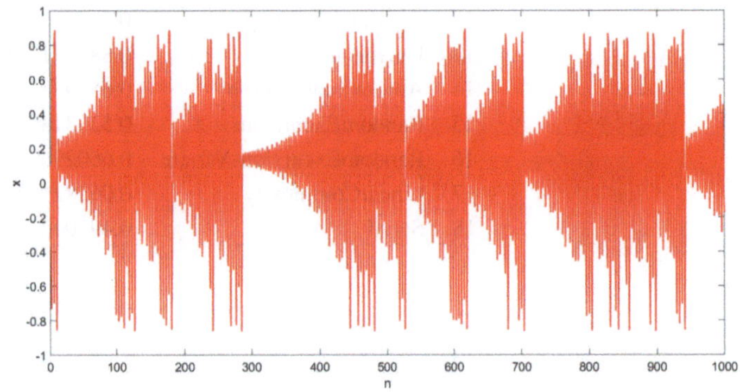

(a)

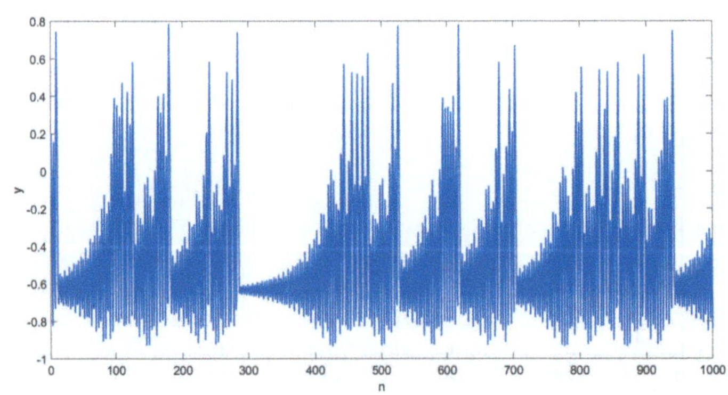

(b)

Fig. 5.4 Chaotic time series of the 2D map, using the coefficient values a_i of the first row in Table 3.2, for the state variables (**a**) x_n and (**b**) y_n

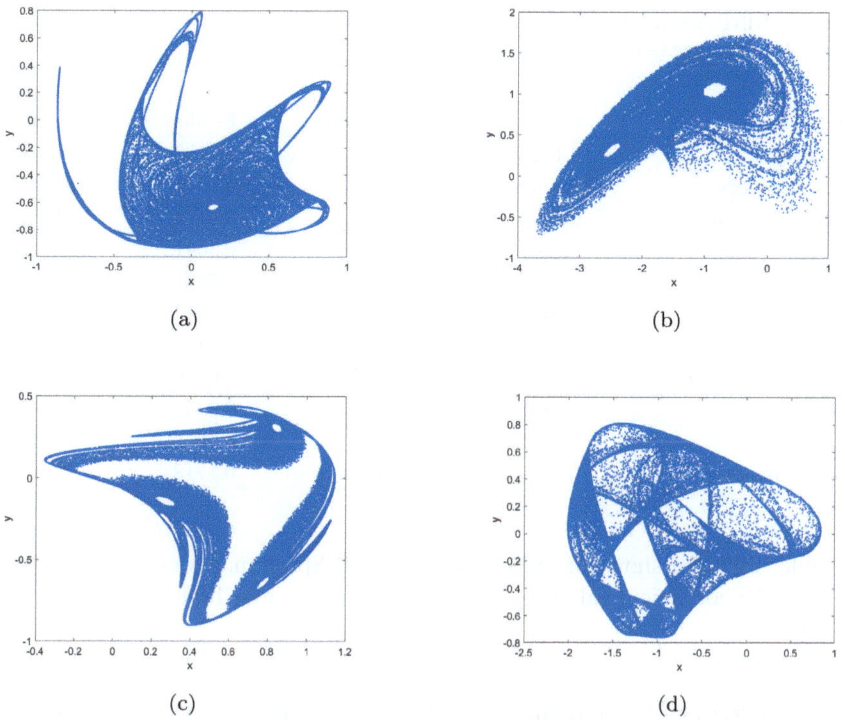

Fig. 5.5 Chaotic attractors for the four sets of coefficients a_i given in Table 3.2: (**a**) Map1, (**b**) Map2, (**c**) Map3, and (**d**) Map4

5.3.1 PRNG from Chaotic Time Series by Applying mod256

The design of a PRNG from chaotic time series can be improved by applying post-processing operations. This section shows the results by performing mod256 operation. The first step in this chapter is the description of the blocks using fixed-point format in either or both a high-level simulation program or Verilog. Using a computer arithmetic of 64 bits, one can take the eight LSBs of each state variable, which can exhibit a random behavior. However, to improve the randomness, the eight LSBs are introduced to the mod256 operation. In a high-level simulation program, this operation implies performing modulo operation between the state variable value and the constant 256, while in a hardware description language such as Verilog, the eight LSBs can be obtained from the bus of each state variable, i.e., by taking the eight LSBs from the 64-bit buses.

Concatenating the eight LSBs, one can perform TestU01 and NIST statistical tests. In such a case, the initial conditions of the state variables are the seed of the PRNG and they are varied within a range where a chaotic dynamics exists. For

Table 5.2 NIST tests results for the case 2D Map4, given in Table 5.2, with initial conditions that are varied in the range $[-0.4, 0.4]$

	x		y	
Statistical test	P-value	Proportion	P-value	Proportion
Frequency	0.1626	99/100	0.3190	99/100
Block frequency	0.5749	99/100	0.4749	100/100
Cumulative sums	0.2896	99/100	0.6371	100/100
Cumulative sums	0.5341	99/100	0.3041	99/100
Runs	0.1223	98/100	0.7981	100/100
Longest run	0.0757	99/100	0.2622	99/100
FFT	0.3669	98/100	0.7597	98/100
Approximate entropy	0.8343	100/100	0.6579	99/100
Serial	0.9878	98/100	0.9240	98/100
Serial	0.5955	97/100	0.3669	98/100
Linear complexity	0.3669	96/100	0.1718	99/100

example, Table 5.2 shows the NIST tests of the 2D Sprott map for case four (Map4), by using the coefficient values a_i given in the fourth row in Table 3.2.

5.3.2 *Verilog Description of a PRNG Based on the 2D Sprott Map*

In Chap. 3, one can see the digital description of the 2D Sprott map, where the design uses computer arithmetic of 64 bits and the architecture is based on two multiplier-accumulator blocks, as shown in Fig. 3.24. A PRNG can be designed by using this 2D map description, where the outputs are fed back to the input to perform iterations in digital hardware. The random binary sequences are then generated by concatenating the eight LSBs of each state variable at each iteration, from the 64-bit buses.

Taking the block description shown in Fig. 3.24, its feedback description is shown in Fig. 5.6. The block Sprott_map_64 has outputs labeled as RX and RY that are connected to multiplexers to feed the data to the pines XN and YN. The multiplexers control the initial conditions for the first iteration, and afterwards, they connect the outputs of the 2D map to the inputs XN and YN. In this case, the flip-flop type-D labeled as FFD has a low state to charge the initial conditions to the 2D map. After one iteration that is performed during nine clock cycles (CCs), the output of FFD changes to a high state that is controlled by signal LFR to enable the ENB pin, so that the 2D map iterates to produce the binary strings. The signal reset RST in FFD is activated to charge new initial conditions. The PRNG is activated through signal STF from block Sprott_map_64, and pin EOF alerts when a new iteration is ready to start. The Verilog description of the block shown in Fig. 5.6 is given in Listing 5.1, where the module is called

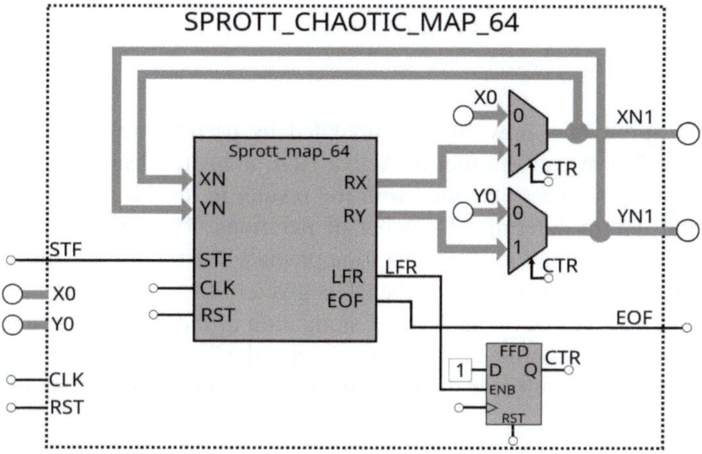

Fig. 5.6 Block diagram description of the 2D Sprott map

SPROTT_CHAOTIC_MAP_64(RST,CLK,STF,X0,Y0,XN1,YN1,EOF), having five inputs and three outputs, with a bus declared as localparam n=64. The Verilog description of the PRNG based on the 2D map instantiates the description Sprott_map_64 U01, given in Listing 3.25.

Listing 5.1 Verilog description of SPROTT_CHAOTIC_MAP_64

```
/*----------------------------------------------------------------
---Project: complete Sprott map                              ---
---Fixed-point format: 1:4:59                                ---
---Inputs: RST - sets the outputs to zero                    ---
           CLK - clock signal                                ---
           STF - start function                              ---
           X0,Y0 - initial conditions                        ---
---Output: XN1,YN1 - output state variables                  ---
           EOF - end of function                             ---
----------------------------------------------------------------*/
module SPROTT_CHAOTIC_MAP_64(RST,CLK,STF,X0,Y0,XN1,YN1,EOF);
    localparam n=64;

    /*FIXED POINT FORMAT: 1 4 59*/
    input RST,CLK,STF;
    input [n-1:0] X0,Y0;
    output [n-1:0] XN1,YN1;
    output EOF;

    wire [n-1:0] RX,RY;
    wire LFR, CTR;

    assign XN1 = (CTR) ? RX:X0;
    assign YN1 = (CTR) ? RY:Y0;

```

```
    Sprott_map_64 U01 (RST,CLK,STF,XN1,YN1,RX,RY,LFR,EOF);
              FFD U02 (RST,CLK,LFR,1'b1,CTR);

endmodule
```

The 2D Sprott map now can be simulated by performing a testbench within Active-HDL. The waveform tool in Active-HDL helps to observe the time series. The testbench generates *.txt files with the results in hexadecimal for each state variable and with a determined number of iterations. The *.txt files can be read using high-level simulation programs to appreciate the time series and attractors. The Verilog code to perform the testbench is given in Listing 5.2. Figure 5.7 shows the time series generated by the Verilog simulation of the 2D map for the first case (Map1), by using the coefficient values a_i given in the first row in Table 3.2. It can be appreciated that the results are in good agreement with the ones shown in Fig. 5.4.

Listing 5.2 Verilog description of the testbench for SPROTT_CHAOTIC_MAP_64 whose description is given in Listing 5.1

```
/*-------------------------------------------------------------
---Project: Testbench for Sprott map                        ---
---Fixed-point format: 1:4:59                               ---
---Inputs:  RST - sets the outputs to zero                  ---
            CLK - clock signal                              ---
            STF - starts a new iteration                    ---
            X0,Y0 - initial conditions                      ---
---Outputs: XN1,YN1 - output state variables                ---
            EOF: end of function                            ---
-------------------------------------------------------------*/
'timescale 1 ns/1 ns
'define ciclos 10
module TB_SPROTT_CHAOTIC_MAP_64();
    parameter ITER = 16'd10000;
    parameter n=64;
    reg RST,CLK,ENB;
    reg [n-1:0] X0,Y0;
    wire [n-1:0] XN1,YN1;
    wire EOF;
    integer file, file1;

   SPROTT_CHAOTIC_MAP_64 DUT (RST,CLK,ENB,X0,Y0,XN1,YN1,EOF);
   always #'ciclos CLK = ~CLK;

   initial
   begin
        CLK=0; RST=1; ENB=1; #0; RST=0;
        X0 = 64'h00CCCCCCCCCCCCD0; //0.1
        Y0 = 64'h01999999999999A0; //0.2
        file  = $fopen("SPR_MAC_Xn1.txt");
        file1 = $fopen("SPR_MAC_Yn1.txt");
        #10;

      for (integer i = 0; i < ITER; i = i + 1)
        begin
```

```
            $fwrite(file,  "%h\n", XN1);
            $fwrite(file1, "%h\n", YN1);
            #180;      // 9 CLK
          end

        $finish;
        $fclose(file);
        $fclose(file1);
    end
endmodule
```

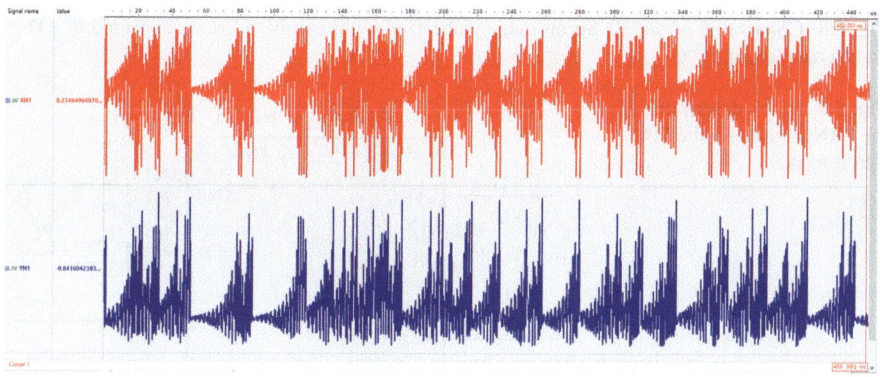

Fig. 5.7 Chaotic time series of the case Map1 by using the coefficient values a_i given in the first row in Table 3.2

5.3.3 PRNG Based on 2D Chaotic Map: FPGA Implementation and Experimental Results

The FPGA synthesis of a chaotic map, as the one shown in Fig. 5.6, can be performed by using the software Quartus. To do this, the experimental time series can be observed in an oscilloscope, by designing additional blocks and by connecting a digital-to-analog (DAC) converter. The process of connecting the FPGA and DAC requires a communication protocol. In this manner, this book uses the DAC8565 having 16 bits. The DAC communicates with the FPGA through the serial peripheral interface (SPI). The connection among the FPGA, DAC, and an oscilloscope is sketched in Fig. 5.8.

As one can infer, the buses in the FPGA have 64-bit length and must be truncated to 16 bits for the connection to the DAC, as sketched in Fig. 5.8, which is not a trivial operation. The truncation process requires to know the maximum and minimum amplitude values of the state variables to be observed in the oscilloscope. The evaluation of these data is useful to establish the number of bits that are assigned to represent the number of bits that will be used for the integer and decimal parts.

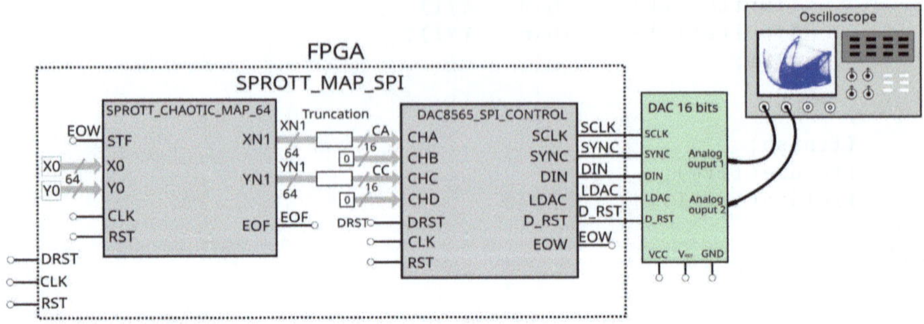

Fig. 5.8 Connection of the 2D Sprott map synthesized within an FPGA and connected to a DAC and an oscilloscope

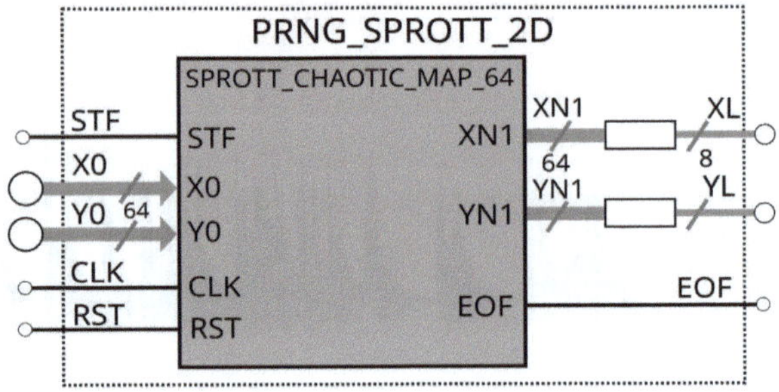

Fig. 5.9 Block description of the PRNG based on the 2D Sprott map

Although the resolution is reduced during the truncation process, one can appreciate the experimental results with good shapes.

As mentioned above, the FPGA implementation of a PRNG requires to take the eight LSBs of the block called SPROTT_CHAOTIC_MAP_64. Concatenating these eight LSBs produced at each iteration in the FPGA generates the chaotic time series. The PRNG based on the 2D Sprott map is shown in Fig. 5.9. In this chapter, the PRNG based on the 2D Sprott map is synthesized within the FPGA Cyclone IV EP4CGX150DF31C7. The hardware resource consumption is given in Table 5.3, where one can also appreciate the throughput and efficiency with a clock of 50 MHz.

The experimental time series of each 2D map case (Map1 to Map4), given in Table 3.2, are shown in Figs. 5.10, 5.11, 5.12, and 5.13, respectively. The time series were observed in a Teledyne Lecroy oscilloscope, and the experimental attractors for the four 2D map cases are shown in Fig. 5.14.

The binary strings can be observed in a logic analyzer. For instance, the behavior of the least significant bit for each state variable can be observed in an oscilloscope, as shown in Fig. 5.15, where one can appreciate the logic levels for each variable x_n and y_n, which are taken from the block PRNG_SPROTT_2D, synthesized into the FPGA Cyclone IV EP4CGX150DF31C7.

Table 5.3 FPGA Cyclone IV EP4CGX150DF31C7 hardware resource consumption for the 2D map PRNG_SPROTT_2D

Resources	Used	Available
Logic elements (LE)	2,056	149,760
Combinational functions	2,005	149,760
Dedicated logic registers	585	149,760
Memory bits	1024	6,635,520
Embedded multiplier 9-bit	96	720
Maximum frequency (MHz)	50.45	–
Clock cycles (MHz)	9	–
Latency at 50 MHz (ns)	180	–
Throughput at 50 MHz (Mbps)	88.8	–
Efficiency (Mbps/LE)	0.043	–

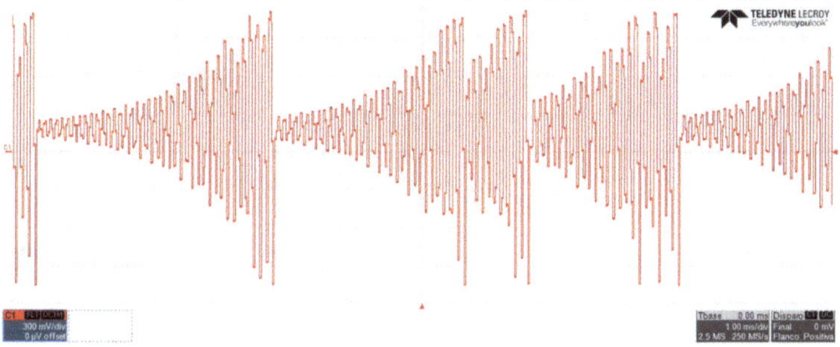

(a)

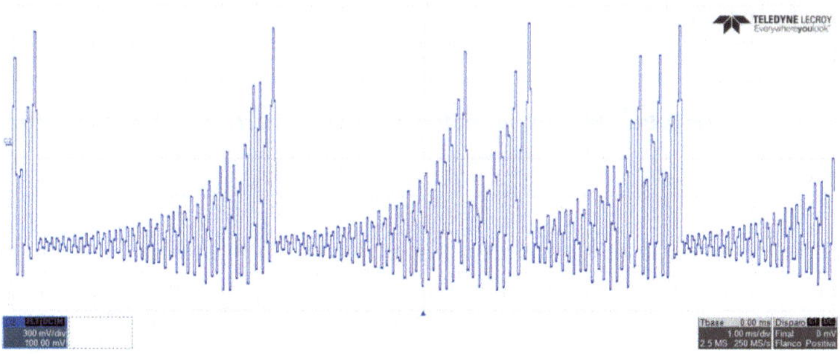

(b)

Fig. 5.10 Experimental time series of the 2D map for case Map1, given in Table 3.2 for the variables (**a**) x_n and (**b**) y_n

5.4 PRNG Based on Chaotic Maps Without Fixed Points

This section shows the design of a PRNG considering three cases of chaotic maps without fixed points and the third one implemented with floating-point numbers.

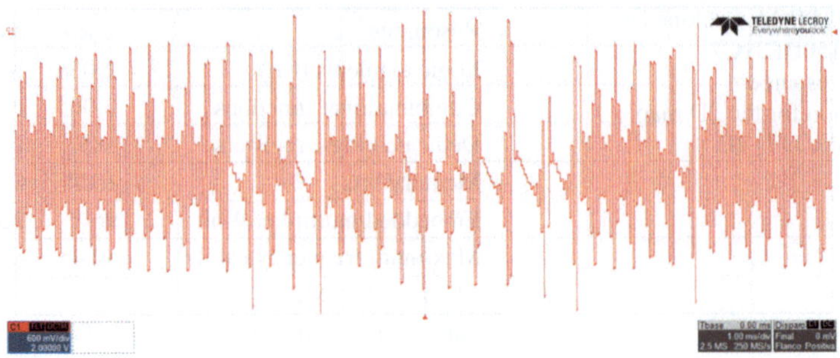

(a)

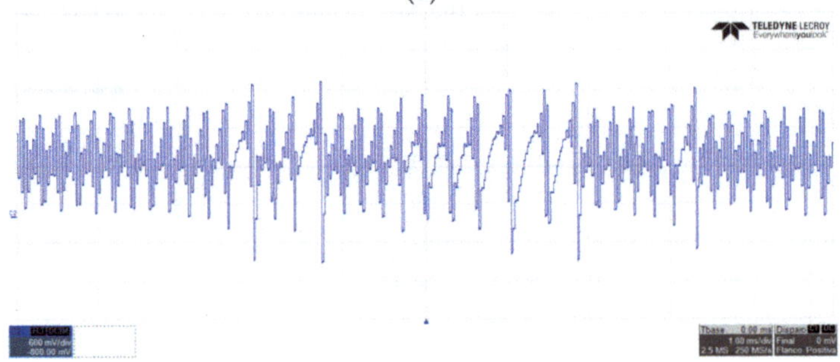

(b)

Fig. 5.11 Experimental time series of the 2D map for case Map2, given in Table 3.2 for the variables **(a)** x_n and **(b)** y_n

5.4.1 PRNG Based on the Chaotic Map Without Fixed Points nfp2

The PRNG described in this subsection uses a map without fixed points, presented in [8] and proposed in [9]. As shown in Sect. 5.2, the operations required to synthesize the map into an FPGA must be simple, like performing additions and multiplications, as done for the 2D Sprott map. The chaotic map described in this part, without fixed points, is slightly simpler than the 2D map [5].

The mathematical description of the chaotic map without fixed points nfp2 presented in [8] is given in Eq. (5.2):

$$\begin{aligned} x(n+1) &= a\,[|y(n)| - 1]\,x(n) + c, \\ y(n+1) &= y(n) + x(n), \end{aligned} \tag{5.2}$$

where the chaotic behavior can be observed by setting the parameter values to $a = 2.3$ and $c = 0.001$. According to [8], this kind of maps without fixed points can be considered more secure because the orbits near the fixed points could not be found. The binary sequences are formed also by concatenating the eight LSBs, as done in

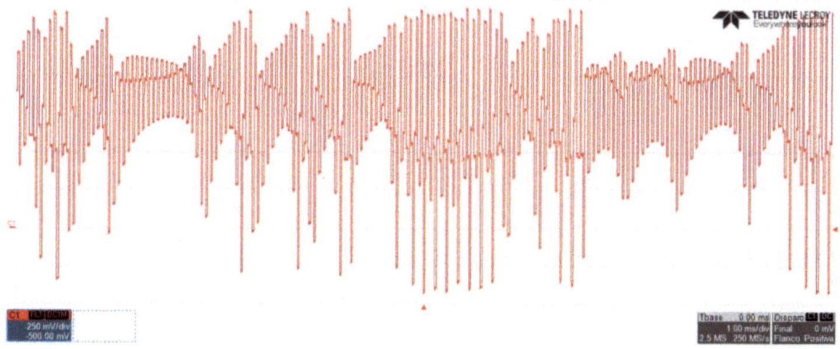

(a)

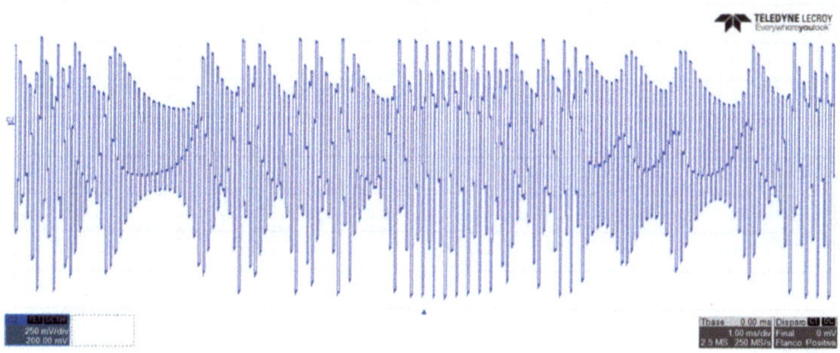

(b)

Fig. 5.12 Experimental time series of the 2D map for case Map3, given in Table 3.2 for the variables (**a**) x_n and (**b**) y_n

[5] for a 2D map. However, a disadvantage of using the map described in Eq. (5.2) is that only one state variable x can be used.

The domain of attraction for the map given in Eq. (5.2) is shown in Fig. 5.16. This figure helps to establish the possible values for the initial conditions $x(0)$ and $y(0)$, which can take any value within the interval $[-0.5, 0.5]$.

For the FPGA implementation of the equations describing the map given in Eq. (5.2) and by using fixed-point arithmetic, one must evaluate the maximum and minimum amplitude values. Taking a look at the domain of attraction shown in Fig. 5.16, the minimum value in the state variable x will be 1.5, and for y the minimum value will be -1. Therefore, the evaluation of the map leads to

$$x(n+1) = 2.3(-1-1)(1.5) = 2.3(-2)(1.5) = -6.9.$$

Thus, $2^3 = 8 > 6.9$, and then 3 bits will be required to represent the integer part. As the computer arithmetic is performed by using 64 bits, then the decimal part will have $63 - 3 = 60$ bits. Therefore, the fixed-point format represents the numbers as 1:3:60.

Different to the FPGA synthesis of the 2D Sprott map, the map without fixed points given in Eq. (5.2) uses an absolute value besides adding and multiplication

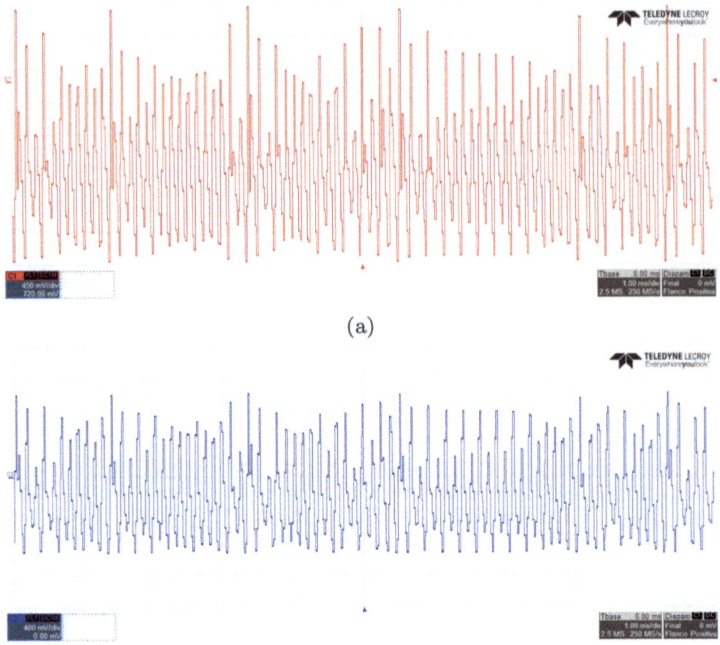

(a)

(b)

Fig. 5.13 Experimental time series of the 2D map for case Map4, given in Table 3.2 for the variables (**a**) x_n and (**b**) y_n

operations. In a 64-bit architecture, a 64-bit variable is a `long` for the C programming language; however, there exists the function `labs()` to calculate the absolute value of a 64-bit number.

The binary sequences are generated then by concatenating the eight LSBs of the output variable $x(n+1)$. Remember that this variable has been coded as a `long` C variable; then its eight LSBs can be taken. To test the randomness of the binary strings, the TestU01 and NIST suite tests were applied to 100 sequences of one million of bits. The results are shown in Table 5.4, where it can be appreciated that all test passed.

Another interesting evaluation is the correlation among three binary sequences, which in this case those sequences that are generated with the smallest change in the initial conditions. In this manner, three binary sequences were generated with initial conditions: [0.0, 0.0] for sequence 1, $[0.0+\epsilon, 0.0]$ for sequence 2, and $[0.0, 0.0+\epsilon]$ for sequence 3, with $\epsilon = 2^{-60}$. The correlation was measured between sequences 1-2, 1-3, and 2-3, and the results are shown in Fig. 5.17. It can be seen that the sequences start to be uncorrelated around byte 30, or around bit 240. This number of bits must be discharged when initial conditions are changed.

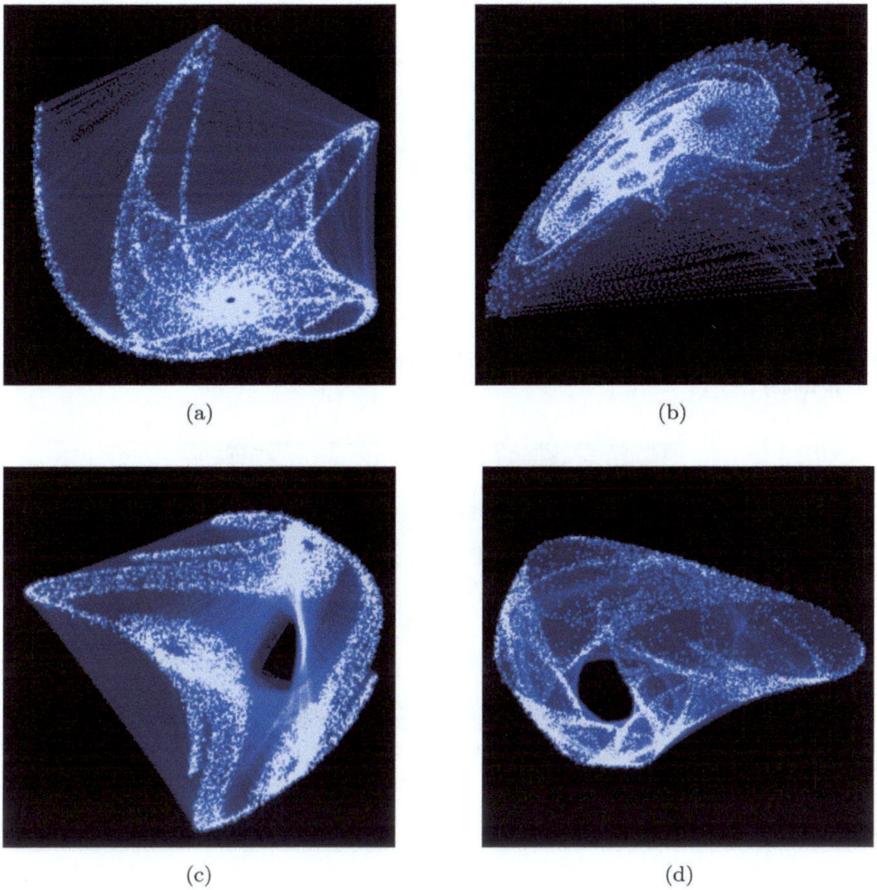

Fig. 5.14 Experimental chaotic attractors of the 2D Sprott map for the four cases given in Table 3.2: (a) Map1, (b) Map2, (c) Map3, and (d) Map4

5.4.2 PRNG Based on the Chaotic Map Without Fixed Points nfp1

Another 2D map is the one known as nfp1, who was presented in [8] and whose mathematical description is given in Eq. (5.3).

$$
\begin{aligned}
x(n+1) &= a\left[y^2(n) - 1\right]x(n) + c, \\
y(n+1) &= x(n) + y(n).
\end{aligned}
\tag{5.3}
$$

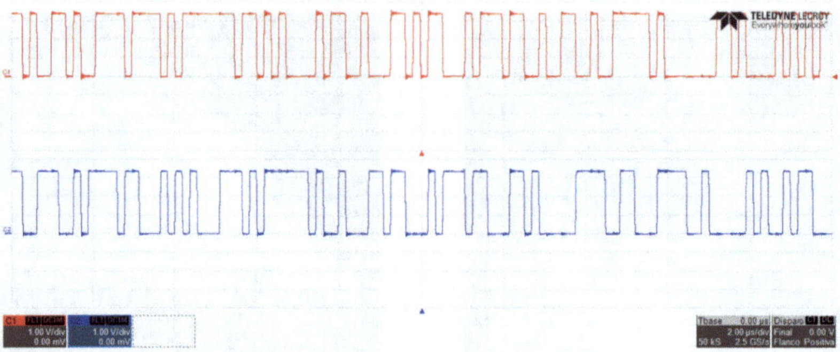

Fig. 5.15 Experimental observation of the LSB of the PRNG based on the 2D Sprott map: x_n (red) and y_n (blue)

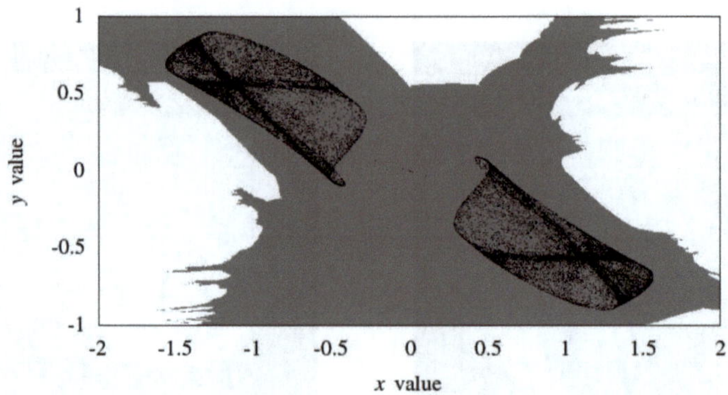

Fig. 5.16 The domain of attraction for the map without fixed points described in Eq. (5.2)

The map nfp1, given in Eq. (5.3), generates chaotic behavior by setting the parameter values to $a = 1.78$ and $c = 0.001$ [8]. One can observe that the difference of nfp1 with the map nfp2 described in the previous Sect. 5.4.1 is that this map uses a square power of $y(n)$ to calculate $x(n+1)$, instead of calculating its absolute values as required in Eq. (5.2). The square power is basically a multiplication and this operation is more complicated than the absolute value.

For the FPGA implementation of the equations describing the map nfp1 given in (5.3), and by using fixed-point arithmetic, one must evaluate the maximum and minimum amplitude values. Figure 5.18 shows the domain of attraction.

To calculate the number of bits to be used in the FPGA implementation with fixed-point arithmetic, the maximum values that can be produced in Eq. (5.3) are obtained by considering $x(n) = -2$, $y(n) = 0$, and $a = 1.78$, so that one gets

$$\begin{aligned} x(n+1) &= 1.78(0^2 - 1)(-2), \\ &= -1.78(-2) = 3.56. \end{aligned}$$

Table 5.4 Results of applying three TestU01 and fifteen NIST statistical tests to 100 sequences of 10^6 bits of the map nfp2. The initial state was set equal to [0.0, 0.0]

	Test name	p-value	Proportion
1	Rabbit	All 40 tests passed	
2	Alphabit	All 17 tests passed	
3	Block Alphabit	All 6 repetitions of Alphabit tests passed	
4	Frequency	0.275709	100
5	BlockFrequency	0.935716	99
6	CumulativeSums	0.527143	100
7	Runs	0.759756	98
8	LongestRun	0.637119	100
9	Rank	0.350485	100
10	FFT	0.108791	99
11	NonOverlappingTemplate	0.515347	99
12	OverlappingTemplate	0.289667	100
13	Universal	0.834308	97
14	ApproximateEntropy	0.474986	99
15	RandomExcursions	0.502645	99
16	RandomExcursionsVariant	0.484376	99
17	LinearComplexity	0.319084	100
18	Serial	0.605927	100

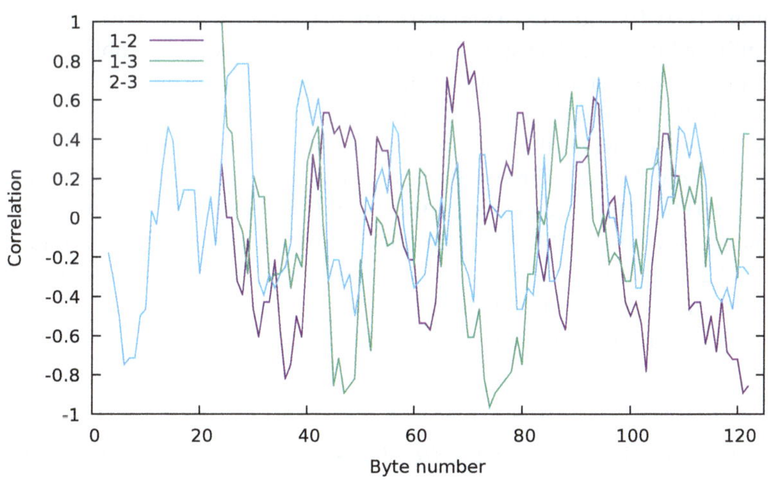

Fig. 5.17 Correlation between sequences 1-2, 1-3, and 2-3 that were generated from the map given in Eq. (5.2) and with three different initial conditions: [0.0, 0.0] for sequence 1, $[0.0 + \epsilon, 0.0]$ for sequence 2, and $[0.0, 0.0 + \epsilon]$ for sequence 3, with $\epsilon = 2^{-60}$

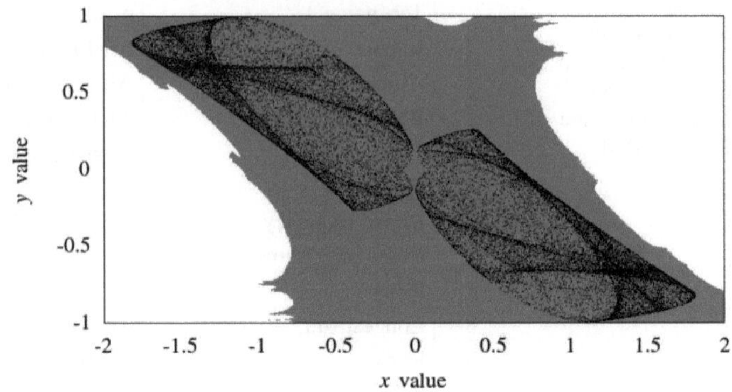

Fig. 5.18 Domain of attraction of the map nfp1 given in Eq. (5.3). This domain was obtained in floating-point arithmetic, considering 40,000 points obtained with initial conditions $(x(0), y(0)) = (-0.5, 0.4)$

Therefore, $2^2 = 4 > 3.56$, and then two bits are required to represent the integer part. For the decimal part, the remaining bits are calculated as $63 - 2 = 61$ bits, and the format is 1:2:61.

The binary sequences are generated by concatenating the eight LSBs of the output variable $x(n+1)$, and the randomness of the generated strings is evaluated by performing TestU01 and NIST suite tests. These tests were applied to 100 sequences of one million bits each. The statistical tests results are shown in Table 5.5, where it can be appreciated that all tests were successfully passed.

Similar to the chaotic map nfp2, the correlation among three generated sequences are now calculated when those sequences are generated with the smallest change in the initial conditions. Three sequences were generated using npf1 with initial conditions: [0.0, 0.0] for sequence 1, $[0.0+\epsilon, 0.0]$ for sequence 2, and $[0.0, 0.0+\epsilon]$ for sequence 3, with $\epsilon = 2^{-61}$. The correlation was measured between sequences 1-2, 1-3, and 2-3, and the results are shown in Fig. 5.19. It can be appreciated that the sequences start to be uncorrelated around byte 25, or around bit 200. This number of bits must be discharged when initial conditions are changed.

5.4.3 PRNG Based on the Chaotic Map Without Fixed Points nfp1 Implemented with Floating-Point Numbers

The previous subsection described the PRNG based on the map nfp1 [8], using fixed-point arithmetic in the format 1:2:61. In this subsection a PRNG is designed with the same map nfp1 but now implemented with floating-point arithmetic. The problem that must be considered from Eq. (5.3) is how to take the least significant bits that do not exist for a floating-point number. In this case, the C function `frexp` can be used. This function converts a floating-point number into normalized

Table 5.5 Results of applying three TestU01 and fifteen NIST statistical tests to 100 sequences of 10^6 bits of the map nfp1. The initial state was set equal to [0.0, 0.0]

	Test name	p-value	Proportion
1	Rabbit	All 40 tests passed	
2	Alphabit	All 17 tests passed	
3	Block Alphabit	All 6 repetitions of Alphabit tests passed	
4	Frequency	0.066882	99
5	BlockFrequency	0.419021	99
6	CumulativeSums	0.600613	99
7	Runs	0.075719	97
8	LongestRun	0.236810	100
9	Rank	0.759756	99
10	FFT	0.574903	99
11	NonOverlappingTemplate	0.475689	99
12	OverlappingTemplate	0.759756	99
13	Universal	0.779188	99
14	ApproximateEntropy	0.883171	98
15	RandomExcursions	0.544384	99
16	RandomExcursionsVariant	0.738194	100
17	LinearComplexity	0.699313	99
18	Serial	0.801459	98

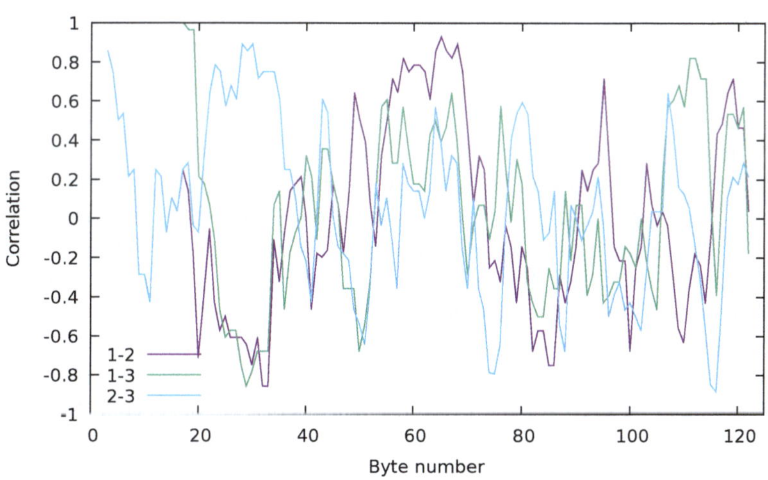

Fig. 5.19 Correlation between sequences 1-2, 1-3, and 2-3, generated from map nfp1 with different initial conditions: [0.0, 0.0] for sequence 1, [0.0+ϵ, 0.0] for sequence 2, and [0.0, 0.0+ϵ] for sequence 3, with $\epsilon = 2^{-61}$

fractional and integral components. The integral component is the exponent and it is an integer number. The fractional component is a floating-point number and is always in the range [0.5, 1). The precision number of bits of a double precision floating-point number (a `double` in C) is 52 bits, and then the given floating-point number $x(n+1)$ in Eq. (5.3) is converted to its normalized fractional and integral components. The normalized fractional component is multiplied by 2^{51} and the resulting number is an integer (a `long` in C), so that the required eight LSBs can be taken to form the binary sequence. In this way, 100 sequences of 10^6 bits were generated and their randomness is evaluated by performing the TestU01 and NIST statistical tests, whose results are shown in Table 5.6. It can also be appreciated that all the generated sequences successfully pass all tests. The PRNG code in C programming language is the following:

```
unsigned char randomMap( )
{
    double xp;
    unsigned long ival;
    int exp;
    double r;

    xp = (yi*yi - 1.0) * xi;
    xp = 1.78 * xp  + 0.0001;
    yi += xi;                  // yi is a global variable
    xi = xp;                   // xi is a global variable

    r = frexp( xi, &exp );
    ival = (unsigned long) (r*2251799813685248); // 2^51
    return( (unsigned char)ival );
}
```

Why use fixed-point arithmetic? Two answers are possible: One is because the implementation in hardware of a PRNG, for example, into an FPGA, must be efficient. The implementation using floating-point arithmetic is hard because it needs to implement a floating-point arithmetic and logic unit which consumes a lot of hardware resources. Instead, the hardware implementation using fixed-point arithmetic needs only the sum and multiplication of integer numbers, which is really easy because those resources already exist in any FPGA.

The second answer is that in fixed-point arithmetic one knows exactly the number of seed that can be used for the PRNG. For example, by observing the domain of attraction of map nfp1 shown in Fig. 5.18, the possible initial values for x_0 and y_0 could be within the interval $[-0.5, 0.5)$. This is with an amplitude equal from 0 to 1. The number $1-2^{-61} = 2^0 - 2^{-61}$ with numbers 2.61 is represented as 2^{61} different values; thus, all the possible values for x_0 and y_0 are $2^{61}2^{61} = 2^{122}$. These values of numbers could be incremented by considering more bits for the fractional part. Such analysis with floating-point numbers is not possible.

Also, the correlation with sequences generated with the minimum represented number is not possible to make it with floating-point numbers. It is not clear which is that minimum number using floating-point numbers.

Table 5.6 Results of applying three TestU01 and fifteen NIST statistical tests to 100 sequences of 10^6 bits of the map nfp1 implemented with floating-point numbers. The initial state was set equal to [0.0, 0.0]

	Test name	p-value	Proportion
1	Rabbit	All 40 tests passed	
2	Alphabit	All 17 tests passed	
3	Block Alphabit	All 6 repetitions of Alphabit tests passed	
4	Frequency	0.224821	100
5	BlockFrequency	0.574903	98
6	CumulativeSums	0.354938	100
7	Runs	0.759756	99
8	LongestRun	0.401199	98
9	Rank	0.275709	100
10	FFT	0.401199	97
11	NonOverlappingTemplate	0.541970	99
12	OverlappingTemplate	0.023545	100
13	Universal	0.935716	99
14	ApproximateEntropy	0.004981	100
15	RandomExcursions	0.712931	99
16	RandomExcursionsVariant	0.359614	99
17	LinearComplexity	0.657933	99
18	Serial	0.730216	99

5.5 PRNG Based on the 3D Lorenz System

The Lorenz system [6, Chs. 4,6] is a continuous-time system that can be described by the ordinary differential equations (ODEs) given in Eq. (5.4), where k is a scale constant:

$$\begin{aligned}\dot{x} &= \sigma(y - x),\\ \dot{y} &= x(\rho - z/k) - y,\\ \dot{z} &= xy/k - \beta z.\end{aligned} \tag{5.4}$$

Chaotic behavior using the Lorenz system can be generated by setting the following values: $\sigma = 10$, $\rho = 28$, and $\beta = 2.6667$; the k value can be set equal to 1. The variation of k to be different of unity is useful to scale the outputs to be within certain interval. The Lorenz system can be simulated by applying the well-known forward Euler algorithm described in Chap. 2 and by using the initial conditions $(x_0, y_0, z_0) = (5.0, 10.0, 15.0)$ and the step size h= 0.001. The discretized equations by applying forward Euler to the ODEs given in Eq. (5.4) lead to Eq. (5.5):

$$\begin{aligned} x_{i+1} &= x_i + h[\sigma(y_i - x_i)], \\ y_{i+1} &= y_i + h[x_i(\rho - z_i/k) - y_i], \\ z_{i+1} &= z_i + h[x_i y_i/k - \beta z_i], \end{aligned} \tag{5.5}$$

The determination of the fixed-point format for the FPGA implementation of the Lorenz system, after evaluating the maximum and minimum amplitude values of the state variables and internal operations, leads to set the format as 1:18:45. Similar to the generation of binary sequences using chaotic maps, in the case of designing a PRNG based on the 3D Lorenz system, it is sufficient to concatenate the eight LSBs of each state variable. In this form, 24 pseudo-random bits can be generated in one iteration of Eq. (5.5). To evaluate the randomness of the binary strings, the TestU01 and NIST statistical tests are performed, whose results on testing 100 sequences of 10^6 bits each one are shown in Table 5.7. It can be appreciated that all tests successfully passed.

To determine which is the minimum value to generate a different binary sequence, four sequences were generated with four different initial conditions of the form: (x_0, y_0, z_0) for sequence 1, $(x_0+\epsilon, y_0, z_0)$ for sequence 2, $(x_0, y_0+\epsilon, z_0)$ for sequence 3, and $(x_0, y_0, z_0+\epsilon)$ for sequence 4, with $\epsilon = 2^{-38}$. The correlation between sequences 1-2, 1-3, 1-4, 2-3, 2-4, and 3-4 was calculated, and the results are shown in Fig. 5.20. The ϵ value was search by hand and it is the minimum to

Table 5.7 Results of applying three TestU01 and fifteen NIST statistical tests to 100 sequences of 10^6 bits generated with the 3D Lorenz system. The initial state was set equal to [5.0, 10.0, 15.0]

	Test name	p-value	Proportion
1	Rabbit	All 40 tests passed	
2	Alphabit	All 17 tests passed	
3	Block Alphabit	All 6 repetitions of Alphabit tests passed	
4	Frequency	0.554420	100
5	BlockFrequency	0.779188	98
6	CumulativeSums	0.798538	100
7	Runs	0.851383	100
8	LongestRun	0.946308	100
9	Rank	0.883171	98
10	FFT	0.275709	99
11	NonOverlappingTemplate	0.472559	99
12	OverlappingTemplate	0.657933	100
13	Universal	0.137282	98
14	ApproximateEntropy	0.071177	99
15	RandomExcursions	0.713988	99
16	RandomExcursionsVariant	0.573941	99
17	LinearComplexity	0.224821	100
18	Serial	0.370700	100

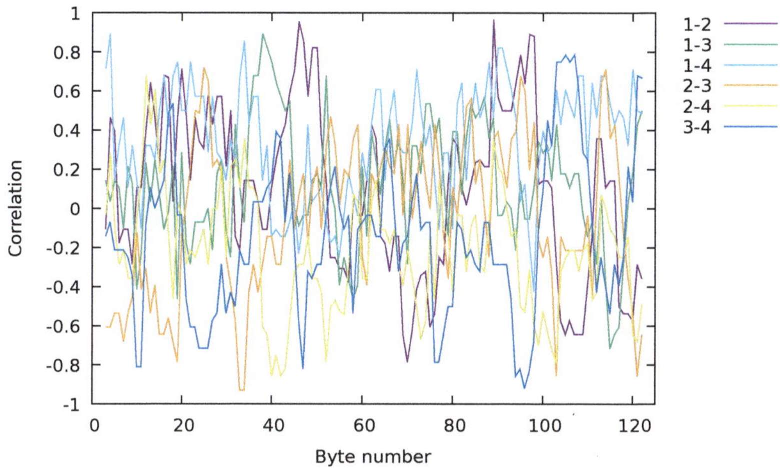

Fig. 5.20 Correlation between sequences 1-2, 1-3, 1-4, 2-3, 2-4, and 3-4, from the generation of four sequences for the Lorenz system by using the initial conditions: [5.0, 10.0, 15.0] for sequence 1, [5.0+ϵ, 10.0, 15.0] for sequence 2, [5.0, 10.0+ϵ, 15.0] for sequence 3, and [5.0, 10.0, +15+ϵ] for sequence 4, with $\epsilon = 2^{-38}$

generate uncorrelated sequences. Note that this value of ϵ is far for the minimum possible value that can be used in the representation that is equal to 2^{-45}.

5.5.1 Verilog Description of a PRNG Based on the 3D Lorenz System

The block description and Verilog code of the 3D Lorenz system is given in Chap. 3, using an architecture of 32 bits. In that chapter, Fig. 3.21 shows the block description for the FPGA implementation. Before synthesizing a PRNG based on the Lorenz system, this chapter shows the synthesis of the 3D Lorenz system with feedback to iterate according to the equations given in (5.5).

The block diagram shown in Fig. 5.21 shows the digital architecture to control the iterations of the Lorenz system, taking as reference the block description shown in Fig. 3.21. In the diagram one can see the block Lorenz_equations_FE_32, whose outputs are labeled as RX, RY, and RZ, and they are connected to PIPO registers that store the results at each iteration. The outputs of each register are connected through multiplexers, to the inputs labeled as XN, YN, and ZN of the block that describes the discretized equations of the Lorenz system. In this architecture, the use of six clock cycles (CCs) is necessary to get the result of a new iteration. A counter is described to count the number of CCs, which is activated after the start signal STF is activated, and when the count reach six, the block activates the signal OPR in order that the three registers store the new value of each state variable. In a similar way, as done

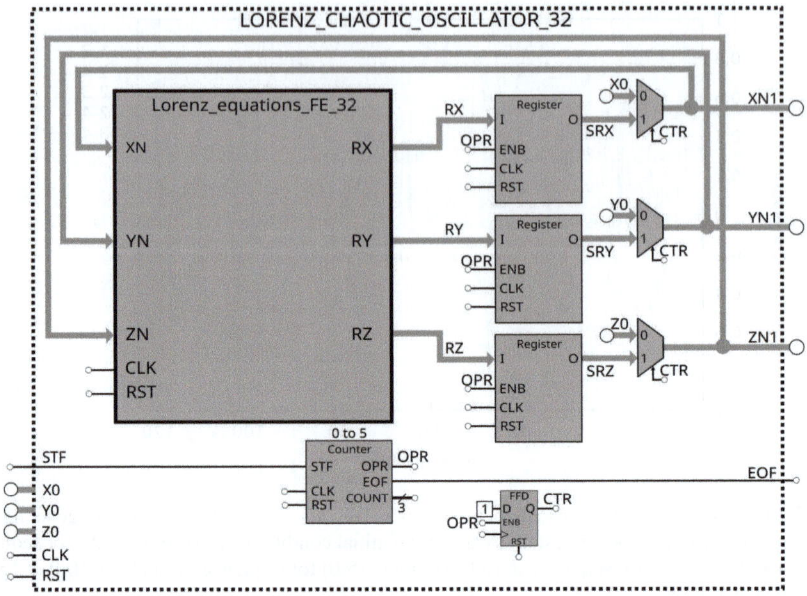

Fig. 5.21 Block description of the 3D Lorenz system to design a PRNG, labeled as LORENZ_CHAOTIC_OSCILLATOR_32

for the 2D Sprott map shown in Fig. 5.6, a flip-flop type-D (FFD) sets the output pin CTR to a low state, allowing the multiplexers to fed the inputs of the chaotic system with the initial conditions, only during the first iteration process. Afterwards, the pin CRT goes to a high state to cancel the transmission of the initial conditions, until the signal reset RST is activated. The pin assigned to the output EOF is used to warn when the system is ready to generate a new iteration. The Verilog description associated to the block shown in Fig. 5.21 is given in Listing 5.3.

Listing 5.3 Verilog description of LORENZ_CHAOTIC_OSCILLATOR_32 shown in Fig. 5.21

```
/*-------------------------------------------------------------
---Project: Complete Lorenz chaotic oscillator             ---
---Fixed-point format: 1:11:20                             ---
---Inputs: RST - sets the outputs to zero                  ---
           CLK - clock signal                              ---
           STF - starts a new iteration                    ---
           X0,Y0,Z0 - initial conditions                   ---
---Outputs:XN1,YN1,ZN1 - output state variables            ---
           EOF: end of function                            ---
-------------------------------------------------------------*/
module LORENZ_CHAOTIC_OSCILLATOR_32
    (RST,CLK,STF,X0,Y0,Z0,XN1,YN1,ZN1,EOF);
    localparam n=32;
    localparam nc=3;
    localparam nclk=6;
```

```

    input RST,CLK,STF;
    input [n-1:0] X0,Y0,Z0;
    output [n-1:0] XN1,YN1,ZN1;
    output EOF;

    wire [n-1:0] RX,RY,RZ;
    wire [n-1:0] SRX,SRY,SRZ;
    wire OPR,CTR;
    wire [nc-1:0] COUNT;

//1.Muxes 2 to 1.
    assign XN1 = (CTR) ? SRX:X0;
    assign YN1 = (CTR) ? SRY:Y0;
    assign ZN1 = (CTR) ? SRZ:Z0;

//2.Module Instantiations.
    Lorenz_equations_FE_32 U01 (RST,CLK,XN1,YN1,ZN1,RX,RY,RZ);
             Register #(n,0) U02 (RST,CLK,OPR,RX,SRX);
             Register #(n,0) U03 (RST,CLK,OPR,RY,SRY);
             Register #(n,0) U04 (RST,CLK,OPR,RZ,SRZ);
         Counter #(nc,nclk) U05 (RST,CLK,STF,COUNT,OPR,EOF);
                        FFD U06 (RST,CLK,OPR,1'b1,CTR);

endmodule
```

The 3D Lorenz system can be simulated by performing a testbench within the software Active-HDL. The simulation results can be saved into a *.txt file to store the results of the state variables. The Verilog description given in Listing 5.4 provides the code to perform the testbench for the 3D Lorenz system, and Fig. 5.22 shows the chaotic time series generated through the waveform tool included into the software Active-HDL.

Listing 5.4 Verilog description to perform a testbench for LORENZ_CHAOTIC_OSCILLATOR_32

```
/*------------------------------------------------------------
---Project: Testbench for Lorenz chaotic oscillator        ---
---Fixed-point format: 1:11:20                             ---
---Inputs: RST - sets the outputs to zero                  ---
           CLK - clock signal                              ---
           STF - starts a new iteration                    ---
           X0,Y0,Z0 - initial conditions                   ---
---Outputs:XN1,YN1,ZN1 - output state variables            ---
           EOF: end of function                            ---
------------------------------------------------------------*/
‘timescale 1 ns/1 ns
‘define ciclos 10
module TB_LORENZ_CHAOTIC_OSCILLATOR_32();
    parameter ITER = 32'd10000;
    parameter n=32;
    reg RST,CLK,STF;
    reg [n-1:0] X0,Y0,Z0;
    wire [n-1:0] XN1,YN1,ZN1;
```

```
     wire EOF;
     integer file, file1, file2;

    LORENZ_CHAOTIC_OSCILLATOR_32 DUT (RST,CLK,STF,X0,Y0,Z0,XN1,YN1
        ,ZN1,EOF);
    always #`ciclos CLK = ~CLK;

    initial
    begin
         CLK=0; RST=1; STF=1; #0; RST=0;
         X0 = 32'h00100000; //1
         Y0 = 32'h00100000; //1
         Z0 = 32'h00100000; //1
         file  = $fopen("Lorenz_32_Xn1.txt");
         file1 = $fopen("Lorenz_32_Yn1.txt");
         file2 = $fopen("Lorenz_32_Zn1.txt");
         #10;

       for (integer i = 0; i < ITER; i = i + 1)
         begin
          $fwrite(file,  "%h\n", XN1);
          $fwrite(file1, "%h\n", YN1);
          $fwrite(file2, "%h\n", ZN1);
          #120; //6 CLK
         end
      $finish;
      $fclose(file);
      $fclose(file1);
      $fclose(file2);
   end

endmodule
```

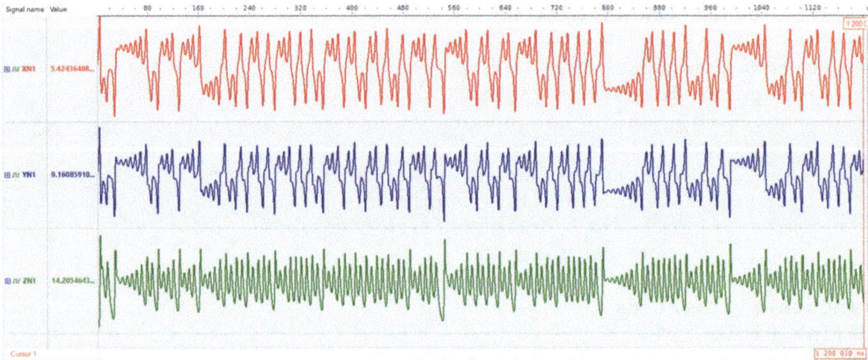

Fig. 5.22 Chaotic time series for each state variable of the 3D Lorenz system observed in Active-HDL

5.5.2 FPGA Implementation and Experimental Results

The FPGA synthesis of the 3D Lorenz system can be performed by using the software Quartus. To observe the chaotic time series and experimental attractors, one can use the architecture shown in Fig. 5.8, where one can see the connection of the outputs provided by an FPGA, a DAC, and an oscilloscope. The block described in this section called LORENZ_CHAOTIC_OSCILLATOR_32 can be used to replace the block of the 2D Sprott map labeled as SPROTT_CHAOTIC_MAP_64. The block of the 3D Lorenz system connects a new state variable to one of the ports of the block called DAC8565_SPI_CONTROL. Similar to the operation performed for the 2D Sprott map, the truncation operation to go from 32 bits to 16 bits, one must evaluate the maximum and minimum amplitude values of the state variables.

The design of a PRNG to be synthesized into an FPGA, and by using the 3D Lorenz system, requires to take the eight LSBs of each state variable. Figure 5.23 shows the block description of the PRNG based on the 3D Lorenz system.

The Verilog description of the PRNG based on the 3D Lorenz system is synthesized into an FPGA Cyclone IV, whose hardware consumption resources are given in Table 5.8. It can be appreciated that a new valid binary string is obtained every 6 clock cycles (CCs), but in this case, the concatenation is performed for the eight LSBs of three state variables x, y, z, which increases the throughput with respect to using the 2D Sprott map, which only has two state variables.

Figure 5.24 shows the experimental chaotic time series of the 3D Lorenz system using a Teledyne Lecroy oscilloscope. Some of the experimental chaotic attractors are shown in Fig. 5.25.

The behavior of the binary strings can be observed in a logic analyzer. Using an oscilloscope, one can also observe the behavior of the LSB of each state variable, as

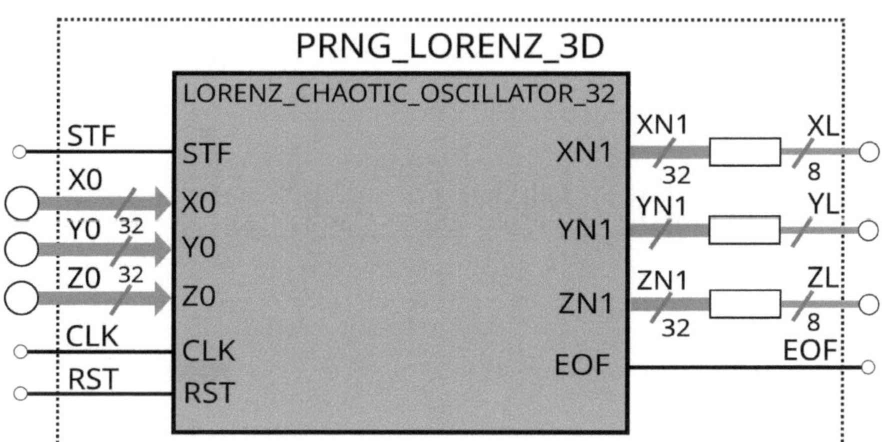

Fig. 5.23 Block description of a PRNG based on the 3D Lorenz system

Table 5.8 FPGA Cyclone IV EP4CGX150DF31C7 hardware resource consumption for PRNG_LORENZ_3D

Resources	Used	Available
Logic elements (LE)	1,113	149,760
Combinational functions	1,110	149,760
Dedicated logic registers	561	149,760
Memory bits	0	6,635,520
Embedded multiplier 9-bit	16	720
Maximum frequency (MHz)	89.77	–
Clock cycles (MHz)	6	–
Latency at 50 MHz (ns)	120	–
Throughput at 50 MHz (Mbps)	200	–
Efficiency (Mbps/LE)	0.179	–

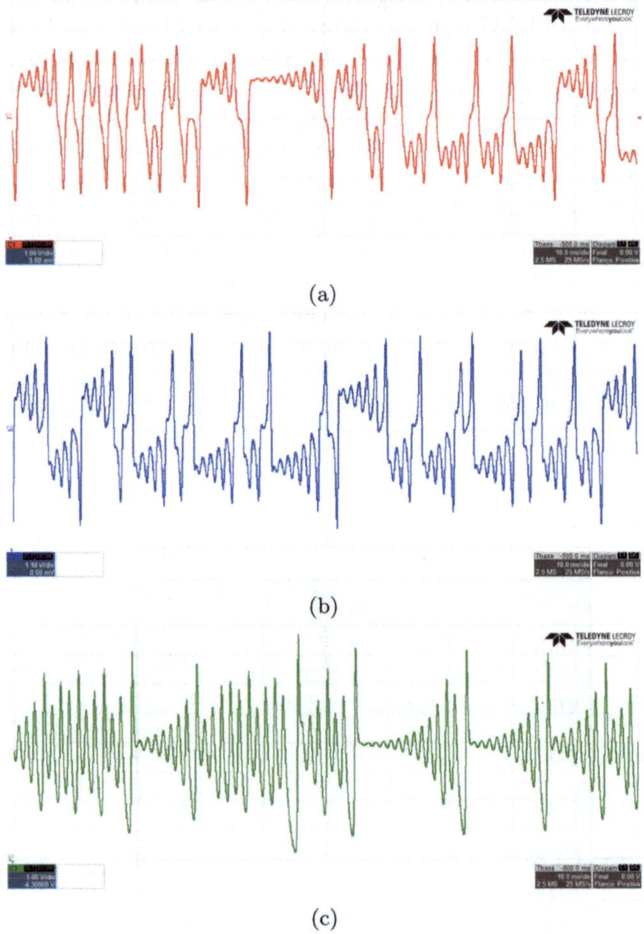

(a)

(b)

(c)

Fig. 5.24 Experimental chaotic time series of the 3D Lorenz system for the state variables: (**a**) x_n, (**b**) y_n, and (**c**) z_n

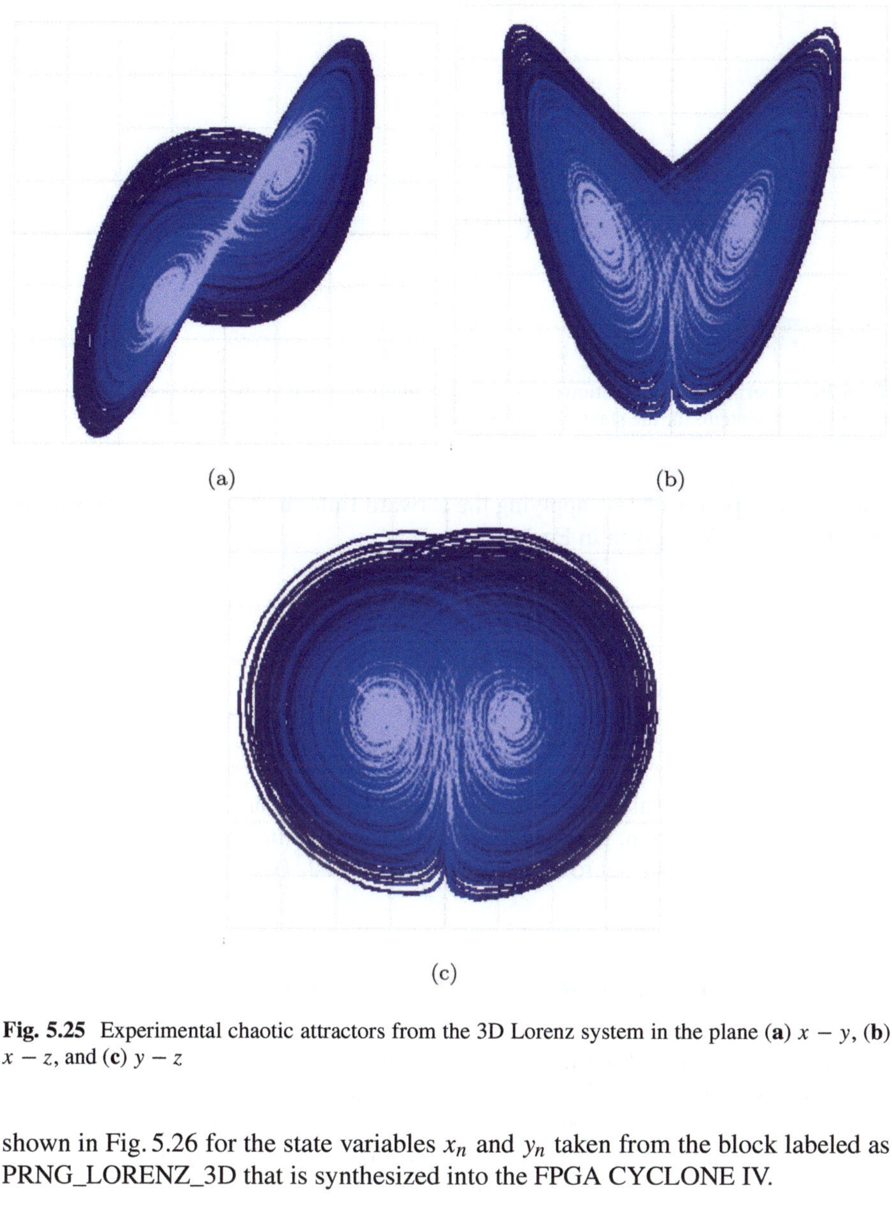

Fig. 5.25 Experimental chaotic attractors from the 3D Lorenz system in the plane (a) $x - y$, (b) $x - z$, and (c) $y - z$

shown in Fig. 5.26 for the state variables x_n and y_n taken from the block labeled as PRNG_LORENZ_3D that is synthesized into the FPGA CYCLONE IV.

5.6 PRNG Based on a 4D Hyperjerk System

In Chap. 2, the mathematical model of a 4D hyperjerk system was given, having four ODEs or state variables, as described by Eq. (2.13). The simulation of this

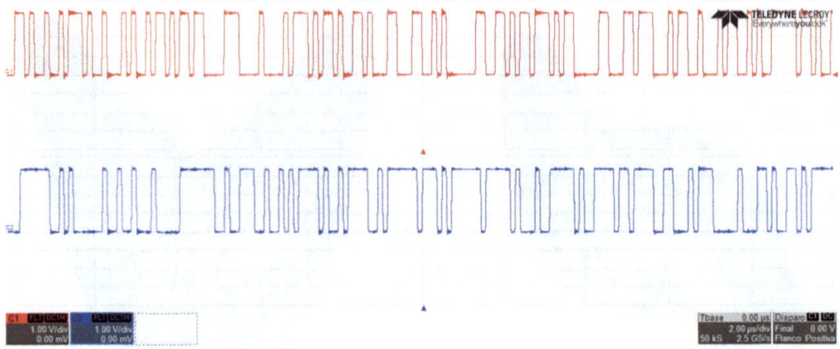

Fig. 5.26 Experimental observation of the binary strings generated by the LSB in the PRNG based on the Lorenz system, for the state variable x_n (red) and y_n (blue)

system can be performed by applying the forward Euler method, which leads to the discretized equations given in Eq. (5.6):

$$\begin{aligned} x_{n+1} &= x_n + h(y_n), \\ y_{n+1} &= y_n + h(z_n), \\ z_{n+1} &= z_n + h(w_n), \\ w_{n+1} &= w_n + h(-x_n - |x_n| - ay_n - bw_n - x_n z_n), \end{aligned} \tag{5.6}$$

where the coefficient values, to generate chaotic behavior, are set to the values $a = 4$, and $b = 2$. In addition, the step size is established to the value of $h = 0.02$, and the initial conditions are set to $(x_0, y_0, z_0, w_0) = (0.2, 0.2, 0.2, 0.2)$.

5.6.1 Time Series, Attractors, and Fixed-Point Format

The simulation of the discretized ODEs given in Eq. (5.6) generates the chaotic times series shown in Fig. 5.27, for the four state variables x_n, y_n, z_n, and w_n, each of them plotted considering 10,000 iterations. To better observe the chaotic attractors, they are shown in Fig. 5.28 considering the data taken during 100,000 iterations.

The amplitudes of the chaotic time series help to find the maximum and minimum amplitude values from all operations in the discretized equations. The fixed-point format is then established as 1:8:55, using 64 bits to perform the FPGA synthesis.

The verification of the randomness of the binary strings generated by concatenating the eight LSBs of the four state variables generates blocks of 32 bits. Generating 100 binary strings of 1,000,000 bit length and after performing NIST tests, the statistical results are given in Table 5.9.

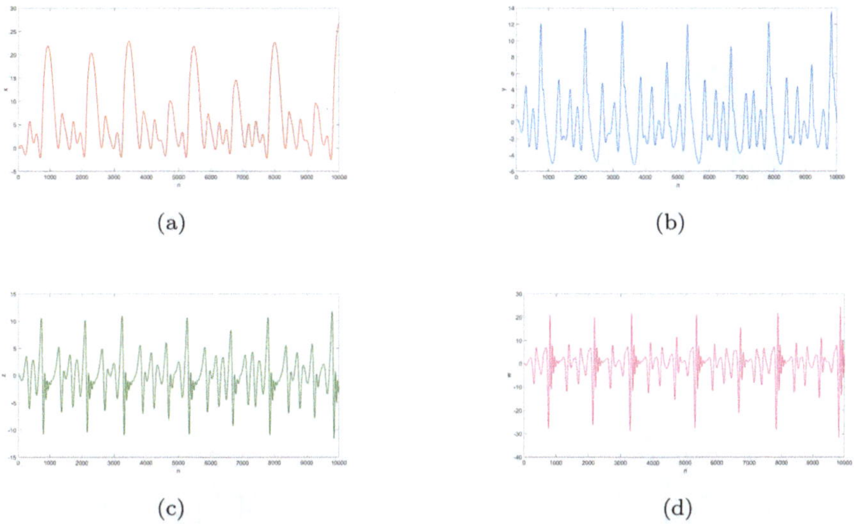

Fig. 5.27 Chaotic time series of the 4D hyperjerk system for the state variable (**a**) x_n, (**b**) y_n, (**c**) z_n, and (**d**) w_n

5.6.2 Block and Verilog Description of a PRNG Based on the 4D Hyperjerk System

The 4D hyperjerk system can be implemented within and architecture of 64 bits to synthesize the equations given in (5.6). As one can see, this system requires the use of adders, subtractors, multipliers, and single constant multipliers (SCMs) to save hardware resources, as described in Chap. 3. Figure 5.29 shows the block description of these equations.

In Fig. 5.29, one can observe the block labeled Subtractor_2N, who performs the subtraction in the form $O = -A - B$. Another special block is the one labeled X_ABS who evaluates the difference between the negative value of the state variable and its absolute value expressed as $-x_n - |x_n|$. In this case, due to the absolute operation, when x_n is negative the difference is zero, but when x_n is positive (or zero), the difference is equal to $-2x_n$. The internal hardware to describe the block X_ABS is shown in Fig. 5.30, who performs a combinational operation to determine the sign of the number to be analyzed, just by verifying the most significative bit (MSB) in the input bus. Depending on the sign, a multiplexer selects one from two results.

All the remaining blocks to synthesize the 4D hyperjerk system given in Eq. (5.6) have been described in Chap. 3. In such a case, the Verilog description requires the design of three SCM blocks in fixed-point format. Therefore, the complete description of the PRNG based on the 4D hyperjerk system requires to synthesize the block diagram shown in Fig. 5.31. The block description is similar to the one

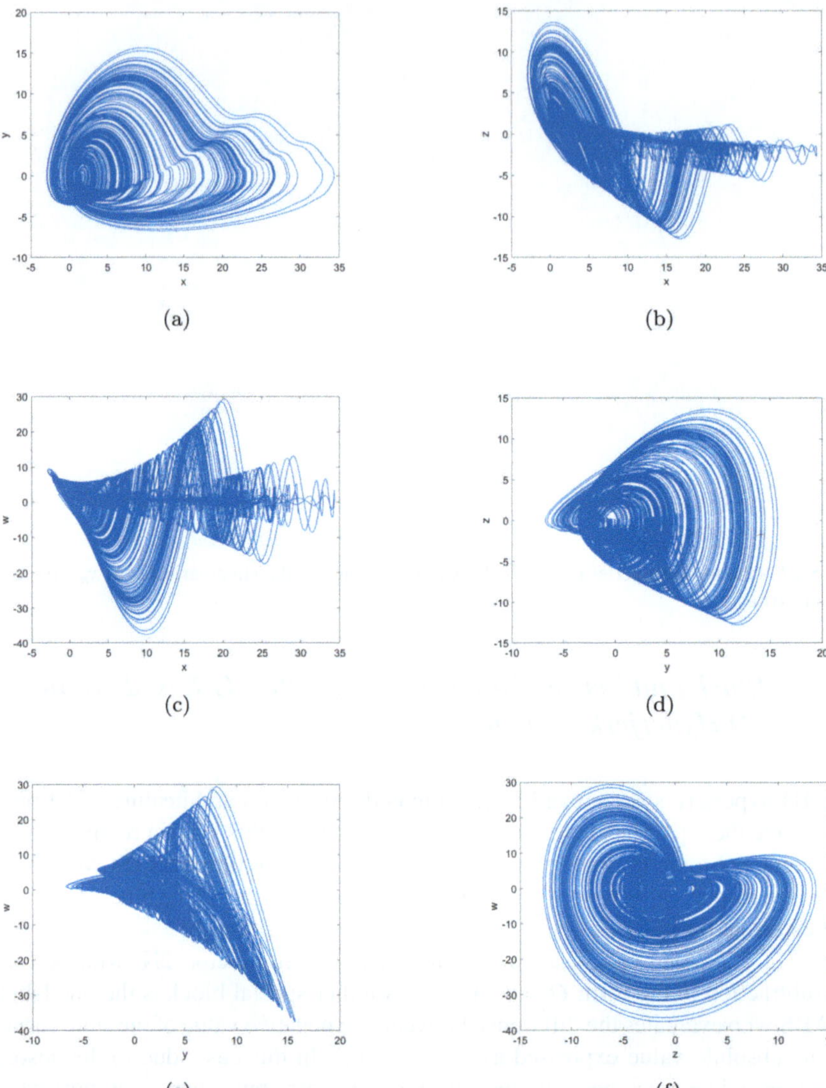

Fig. 5.28 Chaotic attractors from the 4D hyperjerk system in the planes (**a**) $x - y$, (**b**) $x - z$, (**c**) $x - w$, (**d**) $y - z$, (**e**) $y - w$, and (**f**) $z - w$

Table 5.9 NIST statistical tests for the 4D hyperjerk chaotic system, with initial conditions that varied in the range [−0.4, 0.4]

	X		Y		Z		W	
Statistical test	P-value	Proportion	P-value	Proportion	P-value	Proportion	P-value	Proportion
Frequency	0.2622	100/100	0.8831	100/100	0.8676	97/100	0.5544	99/100
Block frequency	0.9357	100/100	0.0216	100/100	0.5955	99/100	0.5749	99/100
Cumulative sums	0.0855	100/100	0.2133	99/100	0.3838	98/100	0.7399	97/100
Cumulative sums	0.1223	100/100	0.1153	99/100	0.2248	98/100	0.4372	99/100
Runs	0.9240	100/100	0.0757	99/100	0.4011	99/100	0.5141	100/100
Longest run	0.1916	98/100	0.2896	98/100	0.6371	100/100	0.9642	98/100
FFT	0.3041	100/100	0.6786	99/100	0.0805	96/100	0.1453	100/100
Approximate entropy	0.0155	100/100	0.6993	100/100	0.2133	100/100	0.9114	98/100
Serial	0.9357	99/100	0.0135	100/100	0.6371	98/100	0.6579	100/100
Serial	0.8343	100/100	0.0456	98/100	0.3504	99/100	0.4011	100/100
Linear complexity	0.3504	100/100	0.8676	96/100	0.3838	98/100	0.3041	99/100

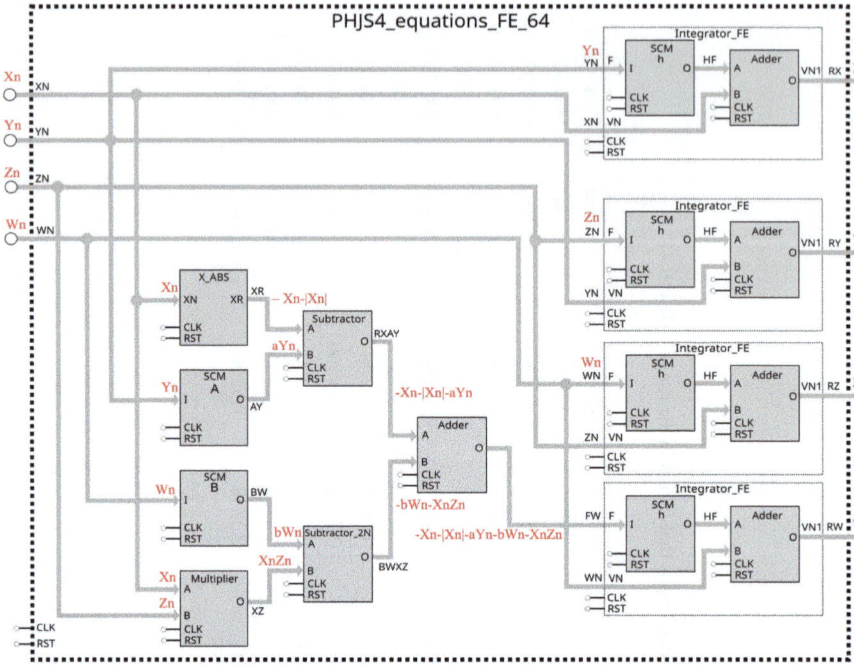

Fig. 5.29 Block diagram description of the 4D hyperjerk system given in Eq. (5.6)

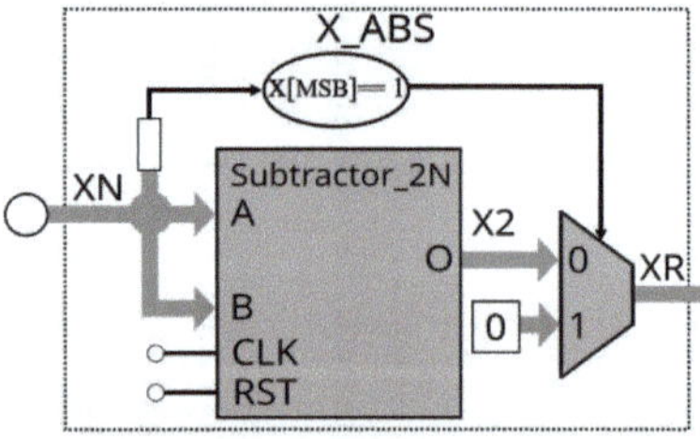

Fig. 5.30 Internal hardware description of the block labeled X_ABS, which evaluates the operation $-x_n - |x_n|$

for the 3D Lorenz system; it just adds another PIPO register to have access to the fourth state variable w_n. The FPGA implementation requires 6 clock cycles (CCs), as for the Lorenz system, to generate new data for each iteration, so that a counter is designed to count from 0 to 5.

The Verilog description of the 4D hyperjerk system shown in Fig. 5.31 follows a very similar structure as the one generated for the 3D Lorenz system given in Listing 5.3.

The block shown in Fig. 5.31 is also simulated in Active-HDL within a testbench with a similar structure as the one given in Listing 5.4 for the Lorenz system. The simulation results of the hardware description is shown in Fig. 5.32, showing the chaotic time series that can be saved into *.txt files.

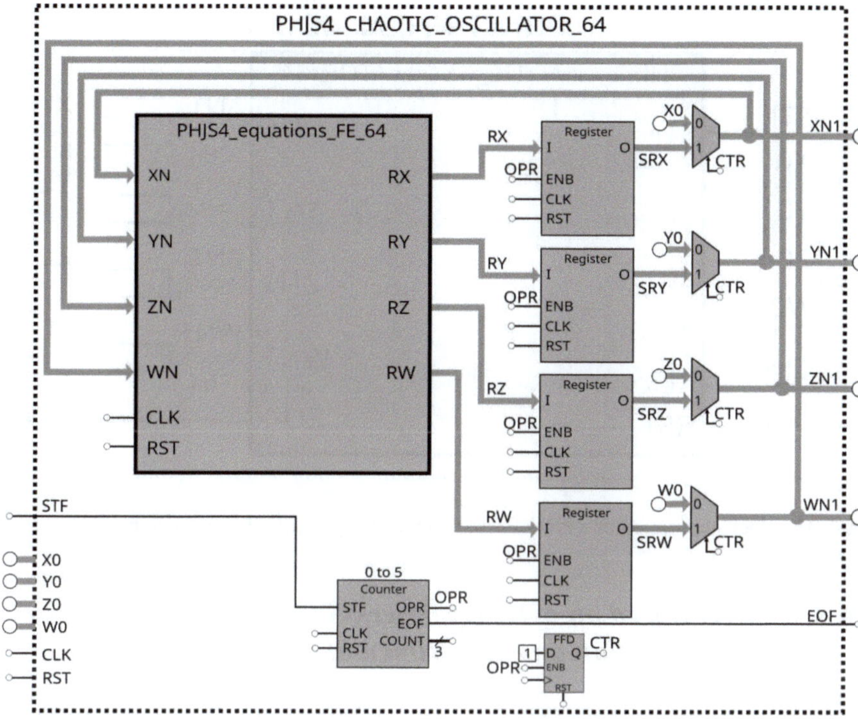

Fig. 5.31 Block diagram description for the 4D hyperchaotic oscillator labeled as PHJS4_CHAOTIC_OSCILLATOR_64

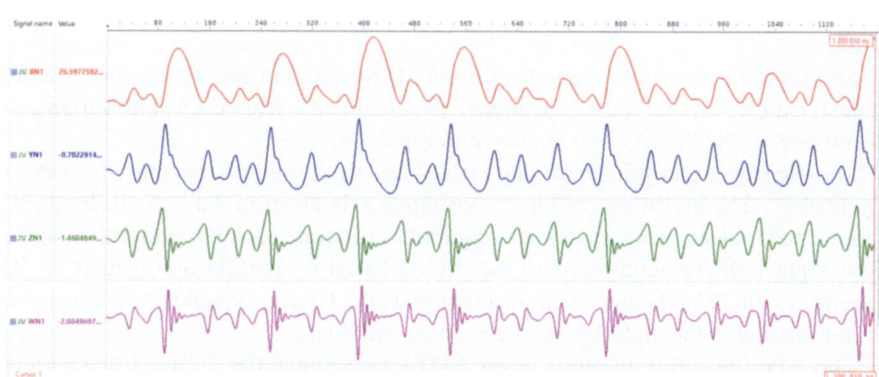

Fig. 5.32 Chaotic time series from the 4D hyperjerk system simulated into Active-HDL

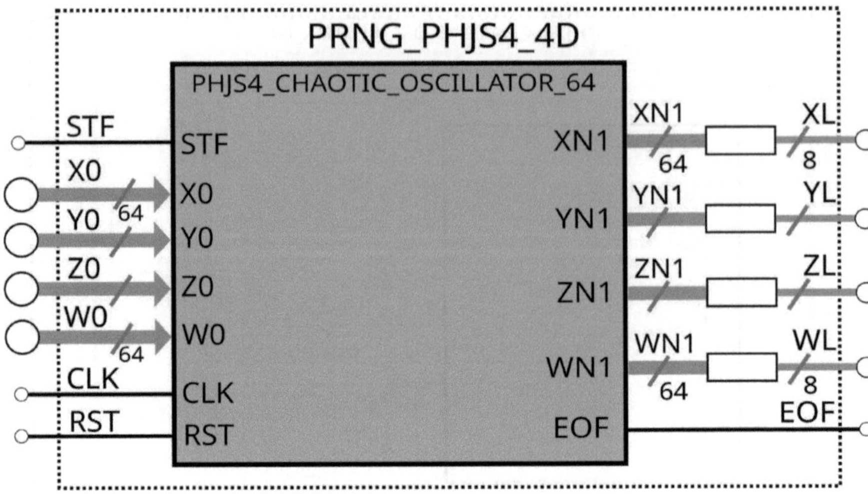

Fig. 5.33 Block description of a PRNG based on the 4D hyperjerk chaotic system

5.6.3 PRNG Based on a 4D Hyperjerk System: FPGA Implementation and Experimental Results

A PRNG based on the 4D hyperjerk system can be synthesized into an FPGA. To observe the experimental chaotic time series and attractors, one must connect the appropriate elements as sketched in Fig. 5.8. In this case, the outputs of the four state variables of the 4D hyperjerk system are connected to the block that drives the DAC.

Similar to the previous examples for the 2D Sprott map and 3D Lorenz system, to design a PRNG, one must concatenate the eight LSBs from each state variable at the outputs of the FPGA. This is shown in Fig. 5.33.

The Verilog descriptions of the PRNG are then synthesized into the FPGA Cyclone IV. The hardware resource consumption is given in Table 5.10. It can be appreciated that six clock cycles are required to generate each iteration but now the throughput is higher compared to the PRNG based on the 2D Sprott map or 3D Lorenz system. This is due to the fact that now, the binary sequences are generated by concatenating the eight LSBs of four state variables.

The experimental time series of the FPGA implementation of the 4D hyperjerk system are shown in Fig. 5.34. They are observed in a Teledyne Lecroy oscilloscope. In addition, Fig. 5.35 shows the experimental observation of the associated chaotic attractors.

The behavior of the binary strings can be observed in a logic analyzer. In this section, Fig. 5.36 shows the behavior of the LSB in an oscilloscope for the state variable x_n and y_n, from the FPGA synthesis of the block labeled as PRNG_PHJS4_4D.

Table 5.10 FPGA Cyclone IV EP4CGX150DF31C7 hardware resource consumption for the block labeled PRNG_PHJS4_4D

Resources	Used	Available
Logic elements (LE)	2,249	149,760
Combinational functions	2,181	149,760
Dedicated logic registers	1,203	149,760
Memory bits	0	6,635,520
Embedded multiplier 9-bit	32	720
Maximum frequency (MHz)	60.68	–
Clock cycles (MHz)	6	–
Latency at 50 MHz (ns)	120	–
Throughput at 50 MHz (Mbps)	266.6	–
Efficiency (Mbps/LE)	0.118	–

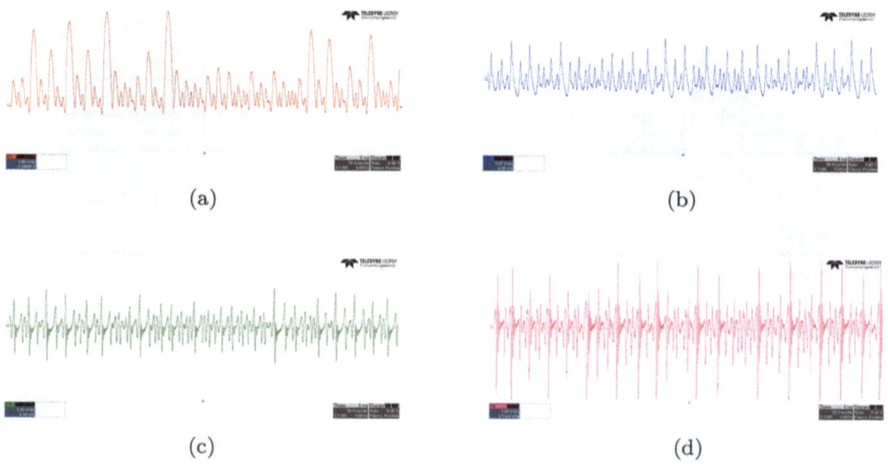

Fig. 5.34 Experimental time series from the FPGA implementation of the 4D hyperjerk system, for the state variable (**a**) x_n, (**b**) y_n, (**c**) z_n, and (**d**) w_n

5.7 PRNG Based on a 5D Chaotic System

The case study of this section is a 5D chaotic system already introduced in [10], which is described by Eq. (2.14), whose discretized equations by applying the forward Euler method are given in Eq. (5.7):

$$
\begin{aligned}
x_{n+1} &= x_n + h(ax_n - y_n z_n - u_n),\\
y_{n+1} &= y_n + h(-by_n + x_n z_n - z_n w_n),\\
z_{n+1} &= z_n + h(-cz_n + x_n y_n),\\
w_{n+1} &= w_n + h(d(y_n - w_n)),\\
u_{n+1} &= u_n + h(ex_n + s),
\end{aligned}
\tag{5.7}
$$

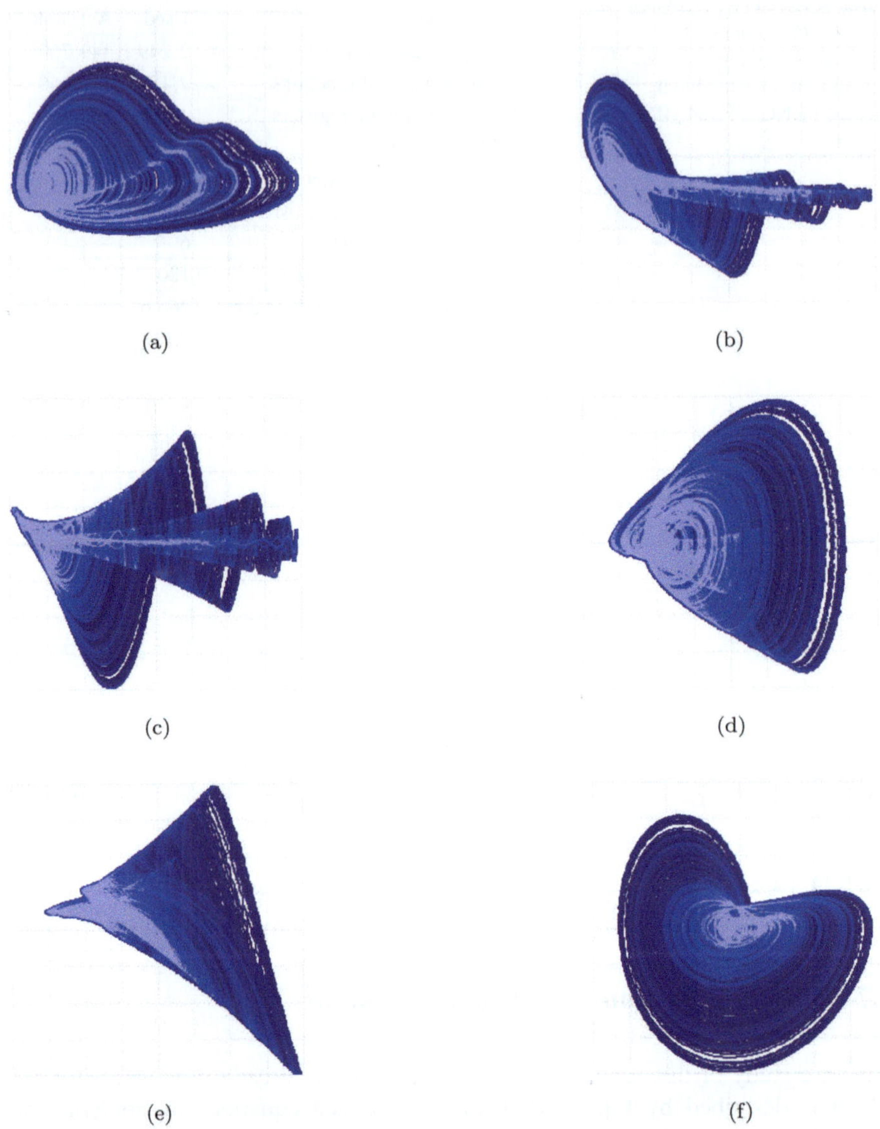

Fig. 5.35 Experimental chaotic attractors from the FPGA implementation of the 4D hyperjerk system, in the plane (**a**) $x - y$, (**b**) $x - z$, (**c**) $x - w$, (**d**) $y - z$, (**e**) $y - w$, and (**f**) $z - w$

where the chaotic behavior is generated by setting the coefficient values to $a = 3.0$, $b = 2.5$, $c = 1.0$, $d = 4.0$, $e = 3.5$, $s = 0.1$ and the step size to $h = 0.001$. The 5D chaotic system is then evaluated by setting the initial conditions to $(x_0, y_0, z_0, w_0, u_0) = (1.0, 1.0, 1.0, 1.0, 1.0)$.

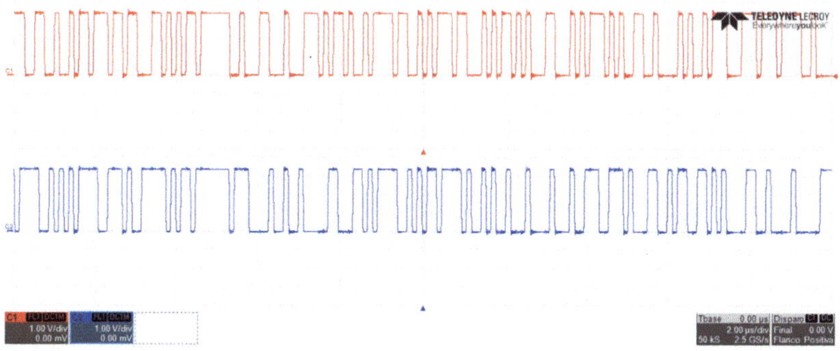

Fig. 5.36 Experimental observation of the behavior of the LSB in the PRNG based on the 4D hyperjerk system, for the state variable x_n (red) and y_n (blue)

5.7.1 Time Series, Attractors, and Fixed-Point Selection

The 5D chaotic system described by five ODEs, as shown in Eq. (2.14), can be simulated into Matlab by using the ODE45 algorithm, and the generated time series must be quite similar to the time series generated by the discretized equations, as the ones given in Eq. (5.7). These simulation operations are performed by using floating-point format, and the results are useful to evaluate the maximum and minimum values that can be generated among all operations, to proceed to the synthesis of the system into an FPGA. Figure 5.37 shows the chaotic time series of the five state variables x_n, y_n, z_n, w_n, and u_n after performing 100,000 iterations. The plotting of these results among the five state variables can generate the chaotic attractors, as shown in Fig. 5.38, which are plotted after evaluating 1,000,000 iterations.

The simulation results of the time series can be used to evaluate the maximum and minimum values that can be generated among all operations in the five equations given in (5.7), as there are some multiplications between two state variables. The values computed in this process are useful to establish the number of bits that are used to perform the Verilog descriptions of the digital blocks to further synthesize the chaotic system into an FPGA. The computer arithmetic is then established with the fixed-point format of 1:8:23, in a total of 32 bits.

5.7.2 Block and Verilog Description of a 5D Chaotic System

As the computer arithmetic is performed by using an architecture of 32 bits, this subsection shows the blocks that synthesize the equations given in (5.7). The blocks that are required to synthesize the equations have already been described in Verilog in Chap. 3. The description of the blocks is shown in Fig. 5.39, where one can

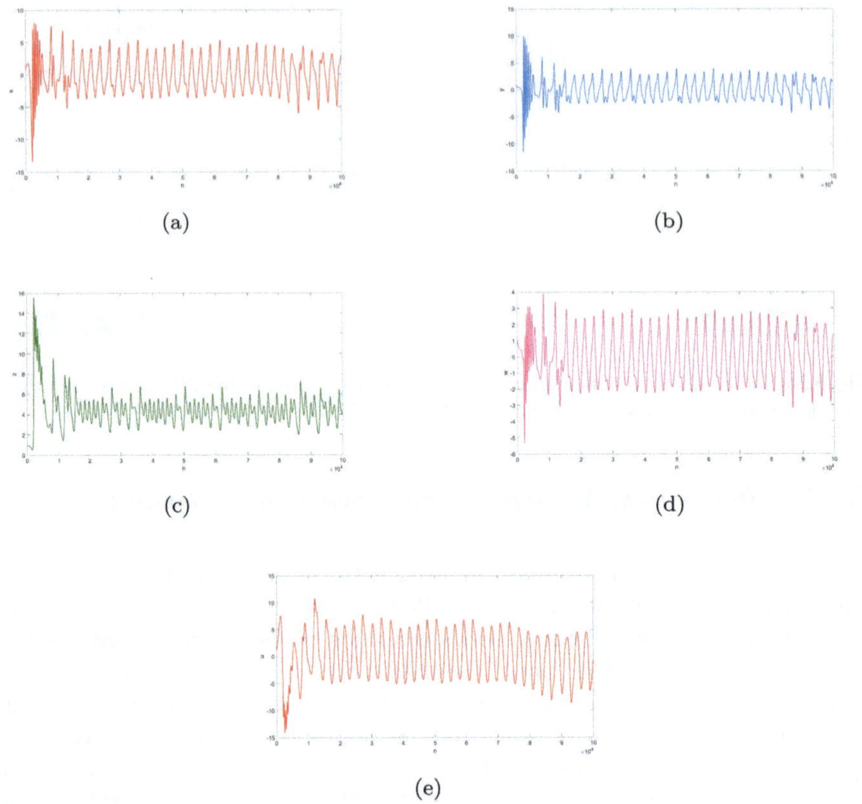

Fig. 5.37 Chaotic time series of the 5D chaotic system, for the state variable (**a**) x_n, (**b**) y_n, (**c**) z_n, (**d**) w_n, and (**e**) u_n

appreciate all the evaluations among the basic blocks until one can get the current values at iteration $n + 1$ for the five state variables.

The arithmetic operations sketched in Fig. 5.39 can be associated to adders, subtractors, multipliers, and single constant multipliers (SCMs). All the remaining blocks have been described in detail in Chap. 3. In that chapter it was mentioned that the SCM blocks must be designed according to the fixed-point format, which in this section is established to have 32 bits.

To synthesize the iterative process of the 5D chaotic system, the outputs must be fed back to the input by designing the appropriate digital control blocks, as shown in Fig. 5.40. This synthesis of the 5D chaotic system requires six clock cycles (CCs) to generate a new iteration in Eq. (5.7).

The block description shown in Fig. 5.40, labeled as SYSTEM5D_CHAOTIC_OSCILLATOR_32, can be described by Verilog as done for the previous 2D Sprott map, 3D Lorenz system, and 4D hyperjerk system. The arithmetic and sequential blocks that are required to synthesize the system into an FPGA have been described

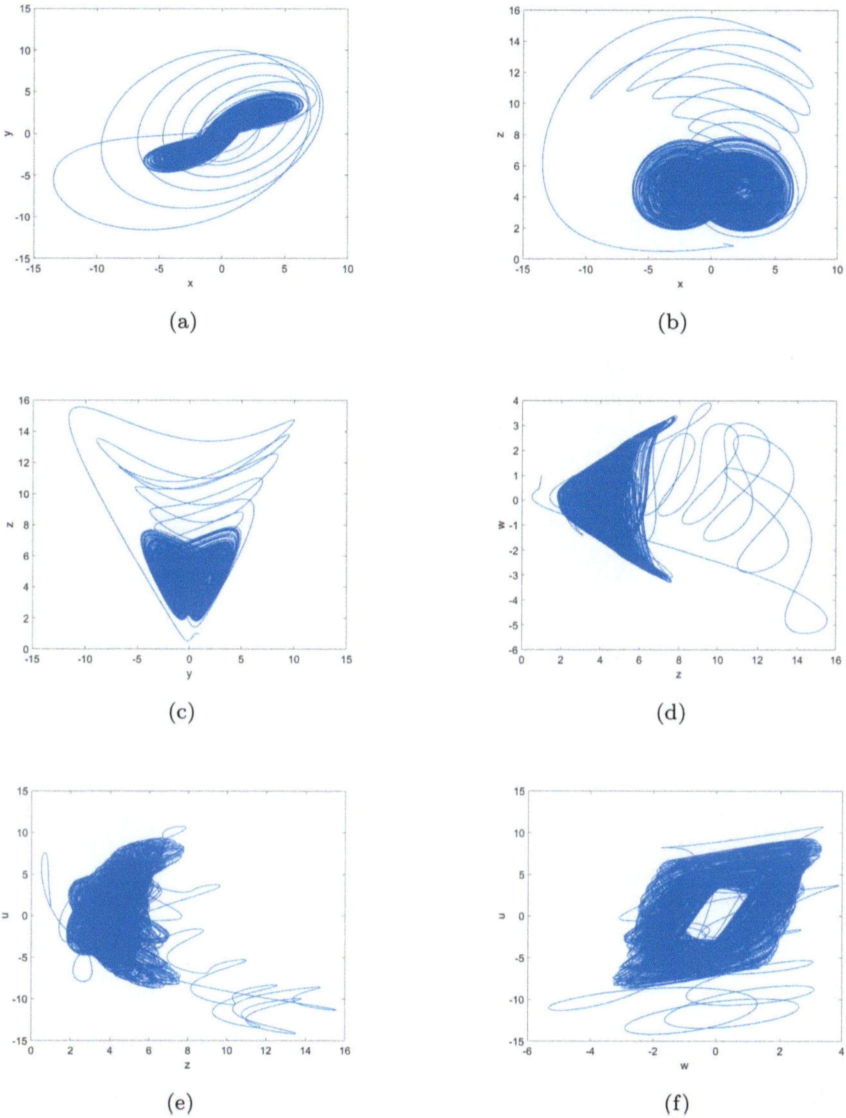

Fig. 5.38 Chaotic attractors generated between two state variables form the time series given in Fig. 5.37, in the plane **(a)** $x - y$, **(b)** $x - z$, **(c)** $y - z$, **(d)** $z - w$, **(e)** $z - u$, and **(f)** $w - u$

and Verilog coded in Chap. 3. The generated Verilog description can be simulated in the software Active-HDL by performing a testbench. The time series obtained by the Verilog code simulation of the 5D chaotic system are shown in Fig. 5.41, where one can see that the results in steady state are quite similar to those obtained by the high-level simulation of Eq. (5.7) and shown in Fig. 5.37. The difference is that

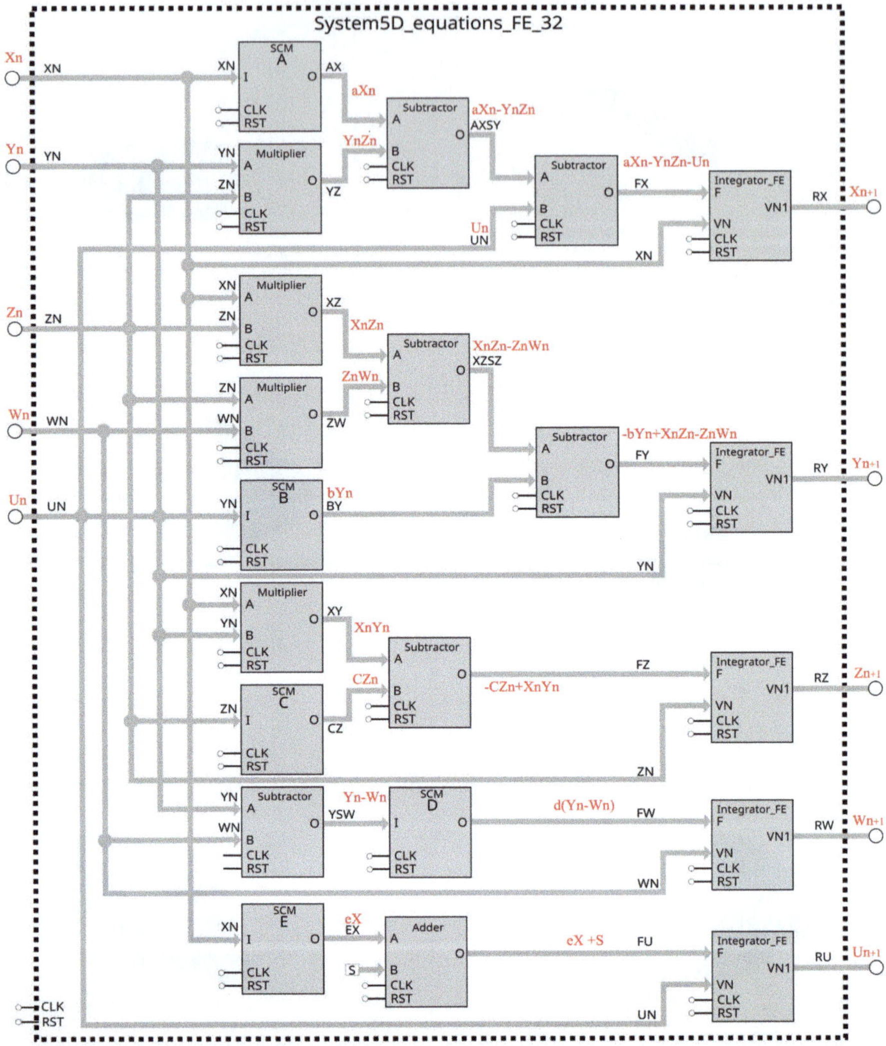

Fig. 5.39 Block description of the 5D chaotic system labeled as System5D_equationsFE_32 synthesized from the equations given in (5.7)

the high-level simulation uses floating numbers and the simulation in fixed-point format uses only 32 bits. Increasing the number of bits can generate chaotic time series more similar to the ones generated by performing high-level simulation using floating-point notation.

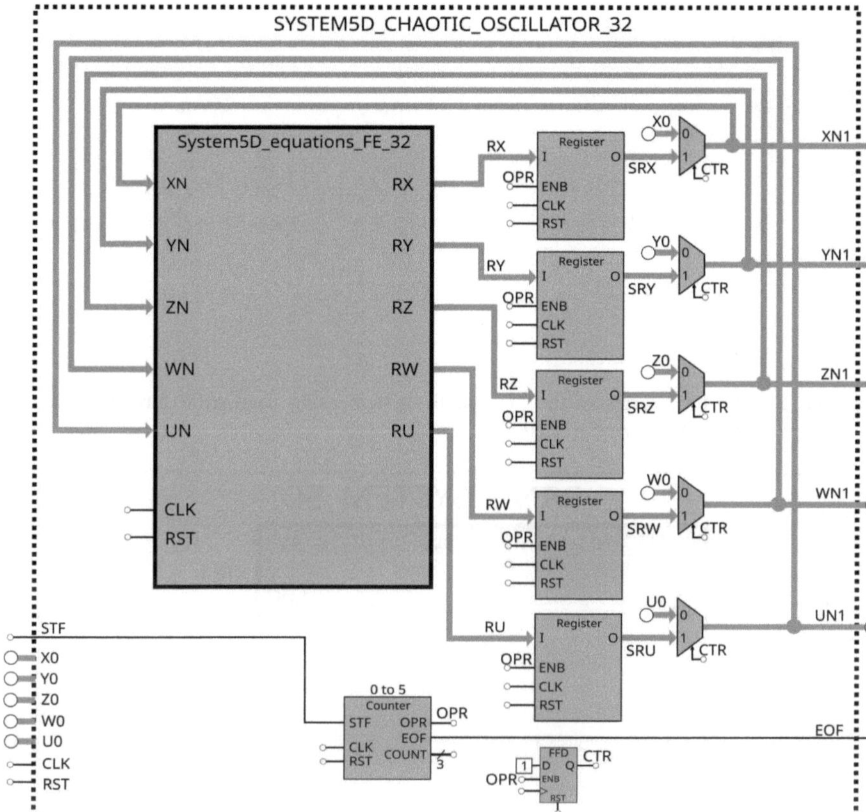

Fig. 5.40 Block description of the 5D chaotic system embedding the block description shown in Fig. 5.39 labeled as System5D_equationsFE_32

5.7.3 PRNG Based on a 5D Chaotic System: FPGA Implementation and Experimental Results

The experimental observation of the chaotic time series and attractors from the FPGA synthesis of the 5D chaotic system can be performed by connecting a DAC and an oscilloscope as shown in Fig. 5.8. In such a case, the 2D Sprott system is replaced by the 5D chaotic system, and in addition, instead of connecting two outputs to the DAC, one must connect now five buses that must be truncated from 32 bits to 16 ones. This is accomplished by adding multiplexers to provide access to the DAC.

The design of a PRNG based on the 5D chaotic system is synthesized by designing the bocks to concatenate the eight LSBs of the five state variables. This process is sketched in Fig. 5.42, where now the effective generated bits are equal to $8 \times 5 = 40$ bits.

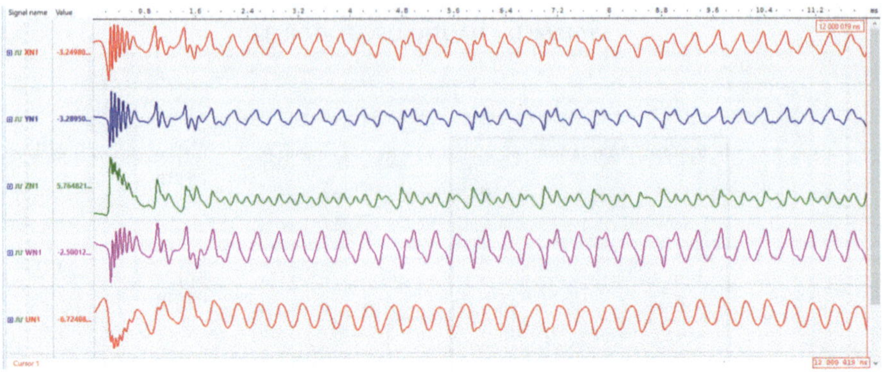

Fig. 5.41 Chaotic time series of the 5D chaotic system using fixed-point format 1:8:23 and simulated with the software Active-HDL

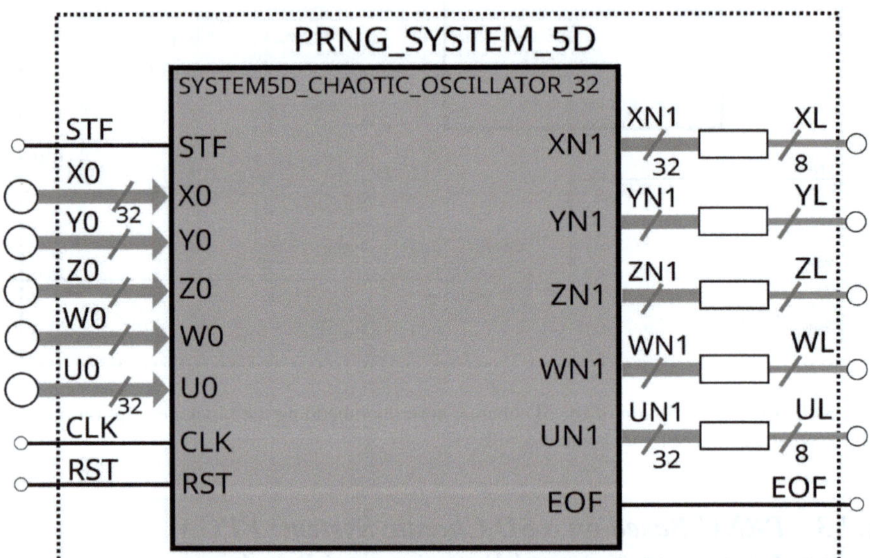

Fig. 5.42 Block description of the PRNG based on the 5D chaotic system

The PRNG based on the block description shown in Fig. 5.42 can be synthesized into the FPGA Cyclone IV EP4CGX150DF31C7, whose hardware resource consumption is given in Table 5.11. It can be appreciated that a new iteration or data is generated every six clock cycles (CCs), but in this case, the PRNG has binary strings from the concatenation of the eight LSBs of five output variables. This clearly improves the throughput compared to using the 2D Sprott map, the 3D Lorenz system, and the 4D hyperjerk system.

The chaotic time series that are observed from the FPGA implementation of the PRNG based on the 5D chaotic system are shown in Fig. 5.43. These data are

Table 5.11 FPGA Cyclone IV EP4CGX150DF31C7 hardware resource consumption for the synthesis of the block labeled as PRNG_SYSTEM_5D

Resources	Used	Available
Logic elements (LE)	1,606	149,760
Combinational functions	1,590	149,760
Dedicated logic registers	947	149,760
Memory bits	0	6,635,520
Embedded multiplier 9-bit	32	720
Maximum frequency (MHz)	79	–
Clock cycles (MHz)	6	–
Latency at 50 MHz (ns)	120	–
Throughput at 50 MHz (Mbps)	333.3	–
Efficiency (Mbps/LE)	0.207	–

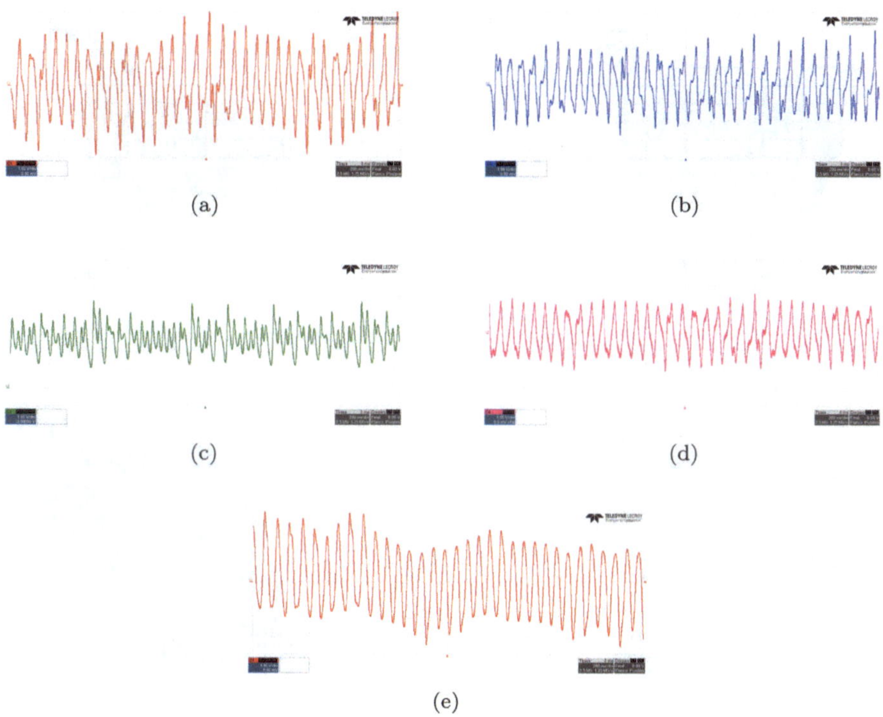

Fig. 5.43 Experimental chaotic time series from the FPGA implementation of the 5D chaotic system, for the state variable **(a)** x_n, **(b)** y_n, **(c)** z_n, **(d)** w_n, and **(e)** u_n

observed in a Teledyne Lecroy oscilloscope, and the portraits between two state variables are shown in Fig. 5.44, where it can be appreciated that the experimental chaotic attractors are quite similar to those obtained from the high-level simulation and the simulation in Active-HDL.

The behavior of the binary strings can be observed in a logic analyzer. Recall that the randomness can be evaluated by performing TestU01 and NIST tests, as shown

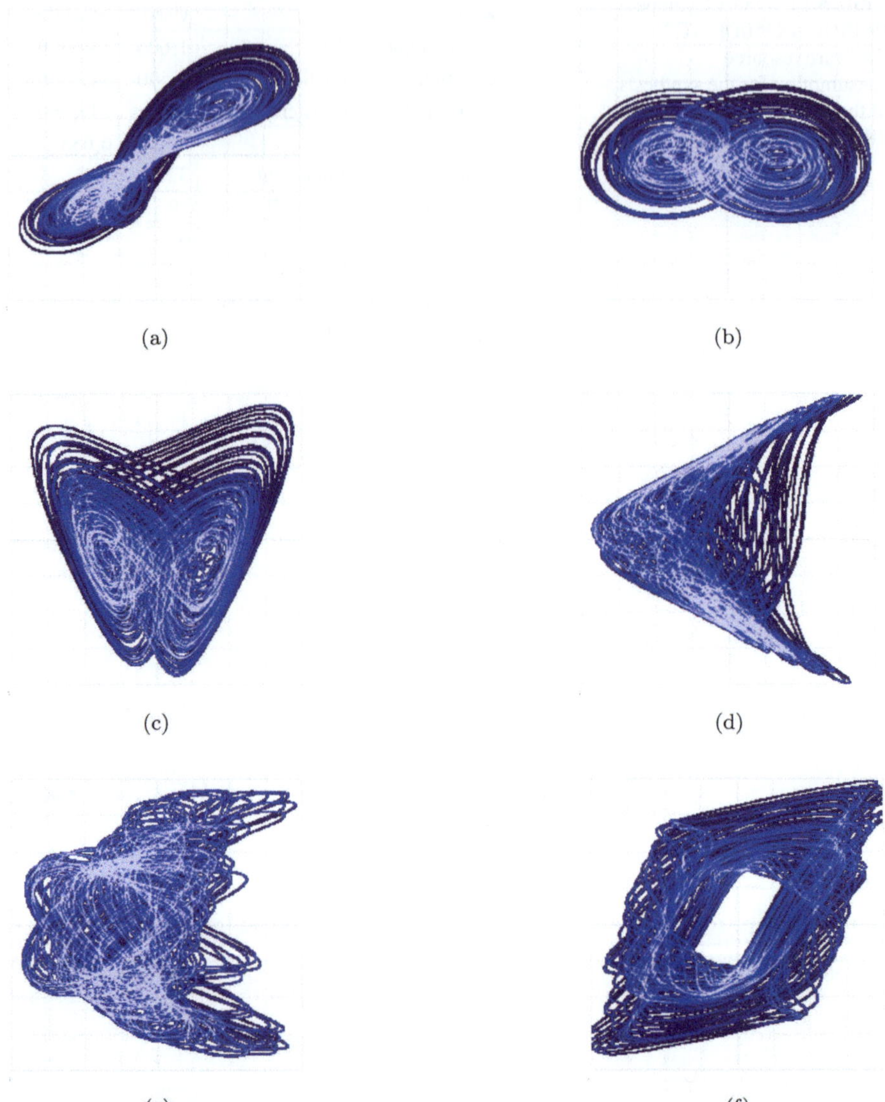

Fig. 5.44 Experimental chaotic attractors from the FPGA implementation of the 5D chaotic system in the plane (**a**) $x - y$, (**b**) $x - z$, (**c**) $y - z$, (**d**) $z - w$, (**e**) $z - u$, and (**f**) $w - u$

in Chap. 4. Besides, the behavior of the LSB can be observed in an oscilloscope, as shown in Fig. 5.45, where the binary strings are observed for the state variables x_n and y_n taken from the block labeled PRNG_SYSTEM_5D that is synthesized in the FPGA Cyclone IV.

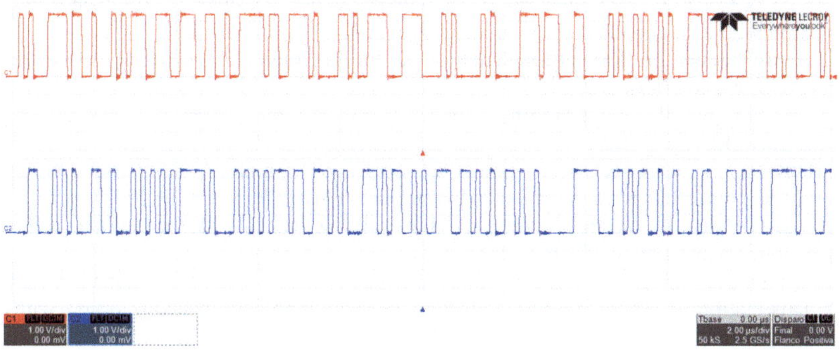

Fig. 5.45 Experimental observation of the binary strings from the LSB of the PRNG based on the 5D chaotic system, for the state variables x_n (red) and y_n (blue)

5.8 Conclusion

This chapter showed the design of PRNGs based on chaotic maps and based on chaotic systems. In the case of using maps, the case study was the 2D Sprott map and others without fixed point. In the case of using chaotic systems, three case studies were considered, namely, the 3D Lorenz system, the 4D hyperjerk system, and the 5D chaotic system. These systems were discretized by applying the forward Euler method. The block descriptions were similar and were based on the blocks described in Chap. 3. All block descriptions for the PRNG based in maps or systems were built with fixed-point arithmetic, except the one using the nfp1 map represented in floating-point notation, as shown in Sect. 5.4.3.

The FPGA implementation of all the maps and systems were detailed and the hardware resource consumption were given in the tables, where it was appreciated that the throughput increases as the number of variables in the map or chaotic system does. This means that the throughput in the PRNG is higher by using the 5D chaotic system, because in such a case, the binary strings are generated by concatenating the eight LSBs of five state variables.

References

1. L.G. de la Fraga, E. Torres-Pérez, E. Tlelo-Cuautle, C. Mancillas-López, Implementation of pseudo-random number generators based on chaotic maps. Nonlinear Dyn. **90**(3), 1661–1670 (2017)
2. T. Stojanovski, J. Pihl, L. Kocarev, Chaos-based random number generators. Part II: practical realization. IEEE Trans. Circuits Syst. I Fundam. Theory Appl. **48**(3), 382–385 (2001)
3. D.E. Knuth, *The Art of Computer Programming Vol. 2, Seminumerical Algorithms*, 3rd edn. (Addison Wesley, Boston, 1998)
4. J.C. Sprott, Automatic generation of strange attractors. Comput. Graph. **17**(3), 325–332 (1993)

5. L.G. de la Fraga, C. Mancillas-López, E. Tlelo-Cuautle, Designing an authenticated hash function with a 2d chaotic map. Nonlinear Dyn. **104**, 4569–4580 (2021)
6. E. Tlelo-Cuautle, J.J. Rangel-Magdaleno, L.G. de la Fraga, *Engineering Applications of FPGAs* (Springer, Berlin, 2016)
7. L.G. de la Fraga, B. Ovilla-Martínez, Generating even more chaotic instances in hardware. Electronics **12**, 332 (2023)
8. J. Ramadoss, A. Ouannas, V.-K. Tamba, G. Grassi, S. Momani, V.-T. Pham, Constructing non-fixed-point maps with memristors. Eur. Phys. J. Plus **137**(211), 1–9 (2022)
9. L.G. de la Fraga, B. Ovilla-Martínez, A chaotic PRNG tested with the heuristic differential evolution. Integration **90**, 22–26 (2023)
10. S.T. Kingni, K. Rajagopal, S. Çiçek, A. Srinivasan, A. Karthikeyan, Dynamic analysis, FPGA implementation, and cryptographic application of an autonomous 5d chaotic system with offset boosting. Front. Inf. Technol. Electr. Eng. **21**, 950–961 (2020)

Chapter 6
PRNG Based on Fractional-Order Chaotic Systems

6.1 Introduction

Chapter 1 shows a description of current works focused on the electronic implementation of chaotic maps, chaotic systems, and fractional-order chaotic systems and their application in the synthesis of pseudo-random number generators (PRNGs). Before the design of a PRNG, one must solve or approximate the solution of chaotic systems modeled by ordinary differential equations (ODEs), or by fractional-order differential equations (FODEs). Some numerical methods to solve ODEs or FODEs are described in Chap. 2, where one can appreciate the discretization of the equations, which are ready to be synthesized by digital hardware using arithmetic, combinational, and sequential blocks, such as adders, subtractors, multipliers, multiplexers, registers, counters, and so on, which are described in Chap. 3. In this manner, Chap. 5 shows the design of PRNGs based on chaotic maps and integer-order chaotic systems. Recall that a PRNG generates random sequences that can be tested by NIST and TestU01 tests, as shown in Chap. 4. Either way, once the randomness of a PRNG is guaranteed, those sequences are ready to develop cryptographic applications, as the ones shown in Chap. 7.

This chapter shows the FPGA synthesis of fractional-order chaotic systems based on memristors, and in addition, the mathematical model includes a delay to improve the complexity in chaotic behavior. The hardware implementation of the FODEs modeling the memristive system with delay is helpful to design a PRNG who can be used to develop applications along machine learning (ML) methods, as shown in [1]. In this article, authors summarize various challenges for the design and implementation of efficient, low-power, and attack-resistant chaos-based encryption schemes with high encryption and decryption rates. Authors also show that the application of ML algorithms can provide solutions to address security challenges in chaos-based encryption methods.

L. G. de la Fraga et al., *Random Number Generators*,
https://doi.org/10.1007/978-3-031-82865-2_6

In the last years, FODEs have been approximated by different numerical methods. The authors in [2] showed that a fractional-order finance system with negative values for system's parameters can decrease the fractional order of a 3D system, as low as 1.74, while the asymptotic stability theorems of the fractional systems are satisfied. The FPGA implementation of discretized equations from FODEs is shown in [3] for different families of fractional-order chaotic oscillators applying the Grünwald-Letnikov method. The analog design of FODEs has also been described in works like [4]. On the side of traditional chaotic systems, the authors in [5] showed the similarity among different sets of FODEs, whose mathematical description led to conclude that they are fractional-order Lorenz-type systems. These Lorenz-type systems and any other system that is modeled by ODEs or FODEs and generates chaotic behavior can be optimized, as shown in [6], and afterwards, they can be implemented by analog or digital hardware.

Nowadays, memristive-based chaotic systems are gaining attention to improve applications, as the design of PRNGs and neural systems. The memristor was introduced by Leon O. Chua in 1971 [7], and in the recent years, memristive-based systems have been implemented with digital hardware, e.g., within FPGAs, as shown in [8], where authors claim an experimental maximum operating frequency of 20 GHz. The authors in [9] show the existence of multistability and initial-offset boosting dynamics in a 3D memristive-based chaotic system. This memristive-based system was implemented into an FPGA and can be extended to synthesize a PRNG. In a similar way, other FPGA implementations of chaotic systems based on memristor are, for example, the hybrid controlled memristor used to design a 5D fractional-order memristor system given in [10] and the 4D hyperchaotic system capable of generating multi-wing chaotic attractors by introducing active magnetron memristors, multi-segmented square functions, and trigonometric functions, given in [11]. The authors in [12] introduced the implementation of a memristive coupled Hopfield neural network (HNN), emphasizing the use of a multistate memristor capable of memorizing multiple conductance states. Difference equations can also be used instead of ODEs and FODEs, as shown in [13], where the authors show the analysis of memristive maps with asymmetry. Recently, the authors in [14] supported the use of memristors to design neural networks in their work titled memristor-induced hyperchaos, multiscroll, and extreme multistability in fractional-order HNN.

All the examples described above, for designs of memristor-based chaotic systems and applications in neural networks, can be improved by adding delay properties, as shown in [15]. In this case, authors claim that fractional-order systems of delay differential equations are very advantageous in analyzing the dynamics of various fields such as population dynamics, neural networking, ecology, and physiology. In addition, memristor-based chaotic systems with time delay can be useful in modern applications combined with ML algorithms, as described in [1]. This chapter shows the design of a PRNG based on a fractional-order memristive system with time delay.

6.2 Block Description of a Fractional-Order 1D Memristive Time-Delay Chaotic System

This section describes the block design of a fractional-order memristive-based chaotic system that includes time delay and is applied to synthesize a PRNG. This system is given in Chap. 2, and the block description can be implemented by using the Verilog codes given in Chap. 3. The block description process detailed in this chapter can be taken as a guideline to describe other fractional-order chaotic systems.

As shown in Chap. 2, the fractional-order 1D memristive time-delay chaotic system can be modeled by Eq. (2.25) [16], where the output $x_{\tau n}$ and the function ψ have different behaviors that are established by the values of n and n_τ, as given by Eqs. (2.26), (2.27) and (2.28). The case study is again given in Eq. (6.1):

$$\begin{cases} i_m = W(\varphi)v_m, \\ W(\varphi) = \gamma - \delta\varphi^2, \\ \dot{\varphi} = v_m - \varphi, \end{cases} \tag{6.1}$$

where $\gamma = 1$ and $\delta = 0.1$. As shown in Chap. 2, this model can be expressed as a first-order nonlinear delay equation, with a delay constant $\tau \in R^+$, as given in Eq. (2.17) and rewritten in Eq. (6.2):

$$\dot{x} = -ax(t) + bW(x_\tau), \tag{6.2}$$

where $W(x_\tau) = \gamma - \delta x_\tau^2$; thus, the integer-order version of Eq. (2.17) can be rewritten by Eq. (2.18), and the FO version can be described by Eq. (2.19), which is given again in Eq. (6.3):

$$D_*^\mu x = -ax(t) + b(\gamma - \delta x_\tau^2), \tag{6.3}$$

which can be solved by applying a numerical method for fractional-order (FO) systems including time delay and by applying the short memory principle described in Sect. 2.2.2, which is useful to save hardware resources. The principle considers the behavior of $f(t)$ only in the values close to the nth iteration, i.e., in the interval $[t - L_m, t]$, and then the short memory length L_m can be associated to Eq. (2.22), so that the number of elements in the summation in Eq. (2.20) is always no greater than $[L_m/h]$. Therefore, if $f(t) \le M$ for $a \le t \le b$, the necessary memory length L_m to achieve the required precision ϵ can be determined through the inequality given in (2.23), which is rewritten in Eq. (6.4):

$$L_m \ge \left(\frac{M}{\epsilon|\Gamma(1-\mu)|}\right). \tag{6.4}$$

The memristor-based chaotic system with time delay generates chaotic behavior by applying the short memory principle by setting $h = 0.01$, $\mu = 0.98$, $\tau = 3.5$,

$L_m = 1.28$, $m = 128$, $n_\tau = 350$ and by considering 128 coefficients ω_j. These coefficients ω_j are computed by substituting m in Eq. (2.21). This is shown in Eq. (6.5):

$$\begin{cases} \omega_0 = \gamma_0 = 1, \\ \omega_j = \gamma_j - \gamma_{j-1}, \ j = 1, 2, ..., m-1, \\ \omega_m = \gamma_{m-1}, \end{cases} \tag{6.5}$$

where $\gamma_j = (j+1)^{1-\mu} - j^{1-\mu}$. In addition, other parameters to be considered in the FPGA implementation are set to $a = 1.0$, $b = 4.5$, $\delta = 0.1$, and $\gamma = 1.0$.

6.2.1 Selection of the Fixed-Point Format

In a similar way, as done for the integer-order chaotic systems described in Chap. 5, the first step is the determination of the maximum and minimum amplitudes of the signals. This allows the assignation of the number of bits that are used to represent the integer part and the number of bits for the decimal part. A simulation of the fractional-order memristive system with time delay given in Eq. (6.3), performing 1,000,000 iterations and using different initial conditions x_0, leads to evaluate all possible maximum amplitudes not only related to the state variable, but also for other multiplications and additions, to give a maximum amplitude in the range ± 32. This chapter shows the use of computer arithmetic with 40 bits, using one bit to represent the sign, 5 bits to represent the integer number, and 34 bits to represent the decimal part, thus obtaining a good resolution. This distribution of the fixed-point format is given in Table 6.1 by using an architecture of 40 bits.

6.2.2 Block Description of the Fractional Memristive System

The selection of the fixed-point format, as the one given in Table 6.1, is the reference point to design the architecture of the system. If one performs an analysis of Eq. (2.25), one can see a summation of 127 products, one ψ function is multiplied by constants, and there is a multiplication between coefficient w_m and the initial condition x_0. Another issue is that function $\Gamma(2-\mu)$, the step size h, and the fractional order μ are constants, causing their values to be previously computed and

Table 6.1 Fixed-point format using 40 bits to synthesize the fractional-order memristive system with time delay

Sign	Integer part	Decimal part
1 bit	5 bits	34 bits

evaluated to work with constants, e.g., K, instead of handling them as parameters. In this manner, Eq. (2.25) can be rewritten by Eq. (6.6):

$$
\begin{aligned}
x_n &= \omega_m x_0 + K\psi(t_{n-1}, x_{n-1}, x_{\tau n-1}) - \sum_{j=1}^{m-1} \omega_j x_{m-j}, \\
x_{\tau n} &= x_{n-n_\tau},
\end{aligned}
\tag{6.6}
$$

where $K = h^{\mu}\Gamma(2-\mu)$.

The block diagram with the highest hierarchy to synthesize the fractional-order memristive system with time delay is shown in Fig. 6.1, which is named FRAC_MEM_TD_EQU, and it fully describes equation (6.6). It can be pointed out that the length of the grey buses is 40 bits, and the black lines represent the control signals. The blocks are labeled in red color with the associated parts of the whole

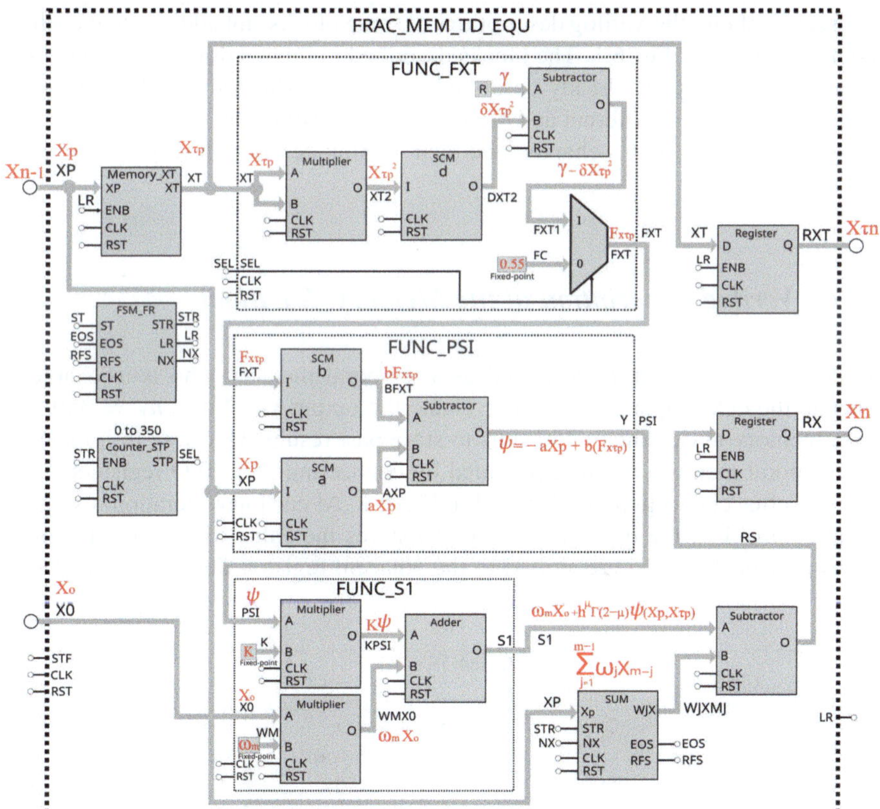

Fig. 6.1 Block diagram of the fractional-order memristive system with time delay by setting $a = 1.0$, $b = 4.5$, $\delta = 0.1$, $\gamma = 1.0$, $K = 0.01084$, and $\omega_m = 0.00017283$

equation, and the black labels represent the I/O ports and wires for the internal connection of the blocks described by Verilog.

The block description of the case study and shown in Fig. 6.1 embeds other labels associated to important pines, as, for example, the input ports are associated to the start control signals (STF), main clock (CLK), and reset (RST). The bus X0 is used to provide and process the initial condition (x_0), and the bus XP is used to process the previous iteration values of x, that is, $X_p = x_{n-1}$. Other buses that are labeled in the figure are the ones associated to RX (x_n) and RXT ($x_{\tau n}$), which are the output buses, and the output pin LR provides a high logic state when the system finishes the current iteration to begin the new one.

6.3 Verilog Descriptions of the Fractional-Order Memristive System with Time Delay

This section shows the Verilog descriptions of some blocks embedded into the whole system described in Fig. 6.1. The blocks described in this section can be composed of basic blocks that are already described in Chap. 3. This section also shows the Verilog description at different and higher levels of abstraction, so that basic blocks are just instantiated in this chapter, and their Verilog descriptions can be revised in Chap. 3.

6.3.1 Verilog Description of the Memory_XT Block

The Memory_XT block is described as a temporal memory who is in charge to generate the delay signal $n - n_\tau$ to evaluate the output $x_{\tau n}$. Since $n_\tau = 350$, this block memory must contain 350 cells to store past results. One way to implement this temporal memory can be performed by connecting 350 PIPO registers, all of them in series connection, as sketched in Fig. 6.2. At each new iteration, using the signal control LR, the register number 350 stores the value of $x_{\tau n}$, and the other values previously stored are shifted to the adjacent register. The output labeled as

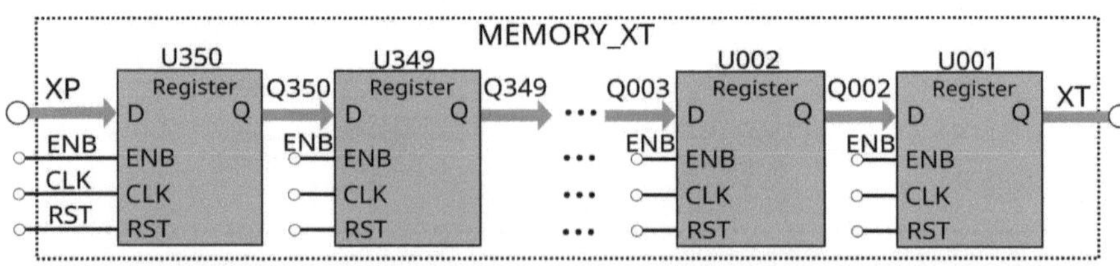

Fig. 6.2 Block diagram of the design of the temporal memory Memory_XT

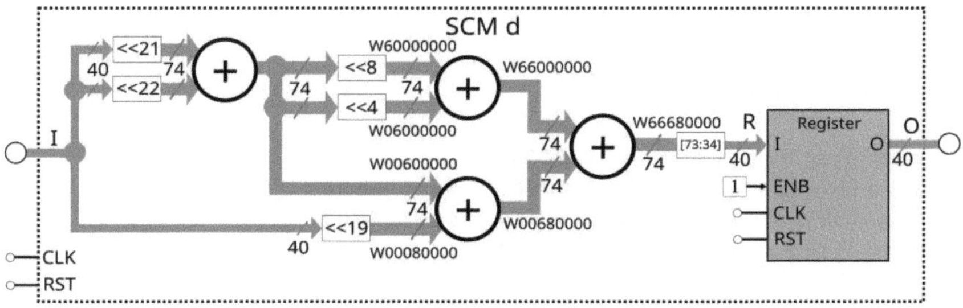

Fig. 6.3 Internal hardware description of the block FUNC_FTX that embeds an SCM_d

XT of this temporal memory is connected to the register number one U001, who provides the required delay. The Verilog description of this temporal memory is too easy by instantiating 350 PIPO registers, and one must describe the connection of each register to another one.

6.3.2 Verilog Description of the FUNC_FXT and Counter_STP Blocks

The block placed between dotted lines called FUNC_FTX describes equation (2.27). This block is in charge of computing the value of $F_{x\tau n}$ from the data input $x_{\tau n}$ and the value at the current iteration n. This block has a multiplexer that is controlled through the signal SEL, which defines if the output takes the value of 0.55 or the one previously defined through the evaluation in Equation $\gamma - \delta x_{\tau n}^2$. In this case, the multiplication by $\delta = 0.1$ can be performed by designing an SCM that is called SCM_d; this is the acronym for single constant multiplier, where δ is the constant value. Figure 6.3 shows the internal hardware description for the constant in its hexadecimal value 0x0066680000, which corresponds to $\delta = 0.100006$ by using a fixed-point format, according to the bit distribution given in Table 6.1. The Verilog description of this SCM is given in Listing 6.1, for the block SCM_d. Recall that these multipliers can be sequential blocks, so that a PIPO register is instantiated through the reserved word Register #(40,0).

Listing 6.1 Verilog description of SCM_d

```
/*-------------------------------------------------------------
---Project: Single Constant Multiplier d=0.100006103515625   ---
---SCM for Fixed-point format: 1:5:34                        ---
---Inputs: RST - sets the output to zero                     ---
           CLK - clock signal                                ---
           I   - logic signed 40-bit input                   ---
---Output: O   - logic 40-bit output                         ---
-------------------------------------------------------------*/
```

```
module SCM_d (RST,CLK,I,O);
    input RST,CLK;
    input signed [39:0] I;
    output [39:0] O;

    wire signed [73:0] W00600000;
    wire signed [73:0] W06000000, W60000000;
    wire signed [73:0] W66000000, W00680000;
    wire signed [73:0] W66680000;
    wire [39:0] R;

    assign W00600000 = (I<<21) + (I<<22);

    assign W06000000 = (W00600000<<4);
    assign W60000000 = (W00600000<<8);

    assign W66000000 = W06000000 + W60000000;
    assign W00680000 = W00600000 + (I<<19);

    assign W66680000 = W66000000 + W00680000;
    assign R = W66680000[73:34];

    Register #(40,0) U01 (RST,CLK,1'b1,R,O);

endmodule
```

Going to Fig. 6.1, one can see that the input SEL is connected to the control pin of a counter with stop, called Counter_STP. This block is shown in Fig. 6.4, and it increments a unity at each iteration n, but when $n > n_\tau$, it stops and activates its output STP, allowing the output of the block FUNC_FXT be different to the constant value 0.55. The Verilog description of Counter_STP block is given in Listing 6.2, while the Verilog description for the block FUNC_FXT is given in Listing 6.3.

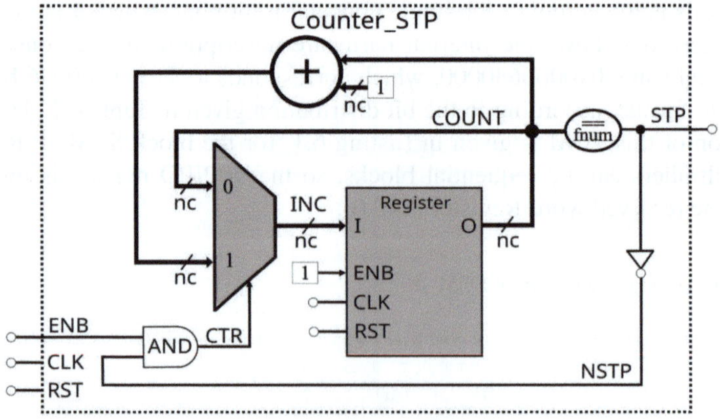

Fig. 6.4 Internal hardware description of the block Counter_STP

Listing 6.2 Verilog description of Counter_STP

```
/*-------------------------------------------------------------
---Project: ascending counter with stop when COUNT == fnum   ---
---Inputs: RST - sets the counter to zero                   ---
           CLK - clock signal                               ---
           ENB - enable counter                             ---
---Output: STP - gives '1' when the count reaches fnum and   ---
                 counter stop increasing until RST signal   ---
-------------------------------------------------------------*/
module Counter_STP #(parameter nc=9,fnum=350) (RST,CLK,ENB,STP);
    input RST,CLK,ENB;
    output STP;

    wire [nc-1:0] INC,COUNT;
    wire NSTP,CTR;

    assign STP = (COUNT==fnum) ? 1'b1: 1'b0;
    assign CTR = ENB & NSTP;
    assign INC = (CTR) ? COUNT + 1'b1: COUNT;
    assign NSTP = ~STP;

    Register #(nc,0) U01 (RST,CLK,1'b1,INC,COUNT);

endmodule
```

Listing 6.3 Verilog description of FUNC_FXT

```
/*-------------------------------------------------------------
---Project: Function FxTau                                   ---
   Fixed-point format: 1:5:34                                ---
---Inputs: RST - sets the output with the initial condition  ---
           CLK - clock signal                                ---
           SEL - select 0.55 or the function value           ---
           XT  - logic input of n bits                       ---
---Output: FXT - logic output of n bits                      ---
-------------------------------------------------------------*/
module FUNC_FXT #(parameter n=40,parameter frac=34)
   (RST,CLK,SEL,XT,FXT);
   input RST,CLK,SEL;
   input [n-1:0] XT;
   output [n-1:0] FXT;

   wire [n-1:0] XT2,DXT2,R,FXT1,FC;

//1. Constants for a length of 40 bits.
   assign  R = 40'h0400000000;  //gamma = 1
   assign FC = 40'h0233333333;  //0.55

   assign FXT = (SEL) ? FXT1: FC;

    Multiplier #(n,frac) U01 (RST,CLK,XT,XT,XT2);
                   SCM_d U02 (RST,CLK,XT2,DXT2);
         Subtractor #(n) U03 (RST,CLK,R,DXT2,FXT1);

endmodule
```

6.3.3 Verilog Description of the FUNC_PSI Block

The block labeled FUNC_PSI, shown between dotted lines in Fig. 6.1, corresponds to Eq. (2.26) that evaluates function $\psi(t_{n-1}, x_{n-1}, x_{\tau n-1})$. In this block the values of the input data are assigned to $F_{x\tau n-1}$ and x_{n-1}. Internally, this block includes two SCMs to describe the constants a and b and includes a subtractor. Since the constant $a = 1$, the description of block SCM_a can be omitted, and therefore, one can directly connect the wire XP with AXP. On the other hand, the design of the block for the constant $b = 4.5$ is labeled as SCM_b, and its internal hardware description is shown in Fig. 6.5. The Verilog description of this SCM_b block is given in Listing 6.4, and the Verilog description for the FUNC_PSI block is given in Listing 6.5.

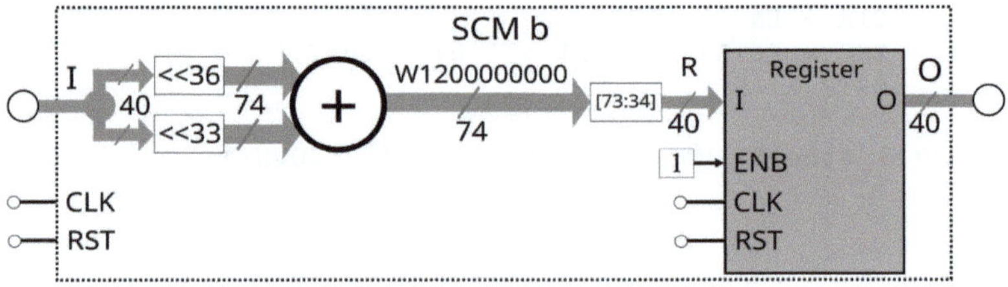

Fig. 6.5 Internal hardware description of the block SCM_b

Listing 6.4 Verilog description of SCM_b

```
/*-----------------------------------------------------------------
---Project: Single Constant Multiplier b=4.5                    ---
---SCM for Fixed-point format: 1:5:34                           ---
---Inputs: RST - sets the output to zero                        ---
           CLK - clock signal                                   ---
           I   - logic signed 40-bit input                      ---
---Output: O   - logic 40-bit output                            ---
-----------------------------------------------------------------*/
module SCM_b (RST,CLK,I,O);
    input RST,CLK;
    input signed [39:0] I;
    output [39:0] O;

    wire signed [73:0] W1200000000;
    wire [39:0] R;

    assign W1200000000 = (I<<36) + (I<<33);
    assign R = W1200000000[73:34];

    Register #(40,0) U01 (RST,CLK,1'b1,R,O);

endmodule
```

Listing 6.5 Verilog description of FUNC_PSI

```
/*-----------------------------------------------------------------
---Project: Function PSI                                        ---
   Fixed-point format: 1:5:34                                   ---
---Inputs: RST - sets the output with the initial condition    ---
           CLK - clock signal                                   ---
           XP,FXT  - logic input of n bits                      ---
---Output: Y - logic output of n bits                           ---
-----------------------------------------------------------------*/
module FUNC_PSI #(parameter n=40) (RST,CLK,XP,FXT,Y);
   input RST,CLK;
   input [n-1:0] XP,FXT;
   output [n-1:0] Y;

   wire [n-1:0] AXP,BFXT;
   assign AXP = XP;   //A=1

              SCM_b U01 (RST,CLK,FXT,BFXT);
    Subtractor #(n) U02 (RST,CLK,BFXT,AXP,Y);

endmodule
```

6.3.4 *Verilog Description of FUNC_S1 Block*

The last block between dotted lines embedded into Fig. 6.1 is called FUNC_S1. This block is in charge of generating the results of the equation $\omega_m x_0 + K\psi(t_{n-1}, x_{n-1}, x_{\tau n-1})$, from which we will require subtracting the value provided by the sum. Although one can design two SCMs for the constants K and ω_m, these values can change if the fractional order μ is changed, so that these constants are handled as parameters and they are used within multiplier blocks instead of SCM. The Verilog description for the FUNC_S1 block is given in Listing 6.6.

Listing 6.6 Verilog description of FUNC_S1 block

```
/*-----------------------------------------------------------------
---Project: Function FxTau                                      ---
   Fixed-point format: 1:5:34                                   ---
---Inputs: RST - sets the output with the initial condition    ---
           CLK - clock signal                                   ---
           PSI - logic input of n bits                          ---
---Output: S1  - logic output of n bits                         ---
-----------------------------------------------------------------*/
module FUNC_S1 #(parameter n=40, parameter frac=34)
   (RST,CLK,PSI,X0,S1);
   input RST,CLK;
   input [n-1:0] PSI,X0;
   output [n-1:0] S1;

   wire [n-1:0] PSI,K,WM,WMX0,KPSI;
```

```
//1. Constants for a length of 40 bits.
    assign  K = 40'h000B1A493F; //0.0108424
    assign WM = 40'h00002D4EC6; //0.000173

    Multiplier #(n,frac) U01 (RST,CLK,PSI,K,KPSI);
    Multiplier #(n,frac) U02 (RST,CLK,X0,WM,WMX0);
              Adder #(n) U03 (RST,CLK,WMX0,KPSI,S1);

endmodule
```

6.3.5 Verilog Description of FSM_FR Block

This block is a finite-state machine (FSM) of Moore type whose main function is devoted to synchronize the functioning of all blocks of the system sketched in Fig. 6.1. The FSM consists of the inputs labeled as ST (start system), EOS (end of sum), RFS (ready for sum), CLK (clock), and RST (reset). The outputs of the FSM are labeled as STR (start sum block), LR (load registers), and NX (store the next x value). The architecture of the FSM is controlled by five states that are shown in the transition diagram shown in Fig. 6.6.

The five states S0 to S4 shown in Fig. 6.6 have the following function:

1. State S0: In this state, the FSM is waiting for the start signal that is provided by the pin ST. In this state S0, the outputs that are initialized to the block SUM update the registers and the memories remain in a low logic state.

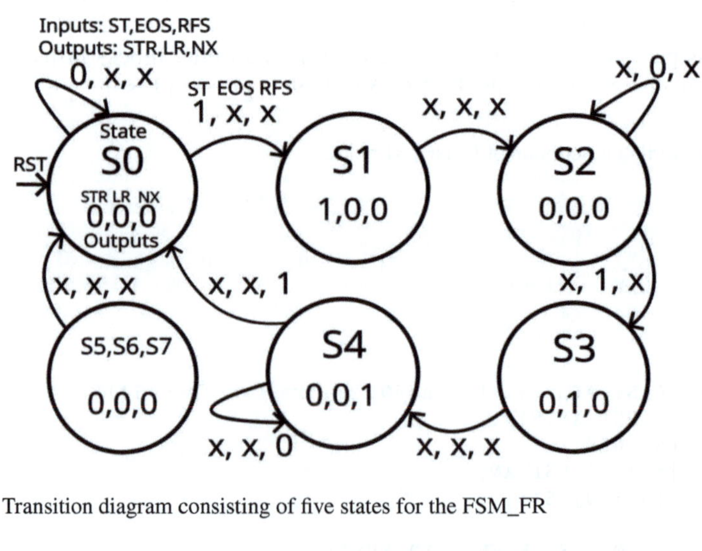

Fig. 6.6 Transition diagram consisting of five states for the FSM_FR

2. State S1: In this state, the FSM activates the control signal STR in order that block SUM begins processing the operation summation $\sum_{j=1}^{m-1} \omega_j x_{m-j}$, and the counter Counter_STP increases its count by a unity.
3. State S2: The FSM waits for the control signal EOS at state S2. This signal is activated once the block SUM ends the summation and provides the final results.
4. State S3: The FSM activates the control signal LR at state S3, to update the block Memory_XT and the output registers for the pines RX and RTX.
5. State S4: In this state, the FSM activates the control signal NX in order that the block SUM updates its short memory, and when this block ends its update, the control signal RFS is activated to return the FSM to state S0.

The transition diagram shown in Fig. 6.6 also includes three states labeled as S5, S6, and S7. However, these states also return the FSM to state S0. The whole digital system that is controlled by the FSM requires 135 clock cycles (CCs) to generate a new value at each output. This process is established considering that the summation operation, performed by the block SUM, requires 128 CCs to provide the result and the remaining 7 CCs are necessary to update the memory blocks and registers. The Verilog description of the FSM is performed by analyzing the transition diagram shown in Fig. 6.6. This process is performed by using the same template detailed in Chap. 3.

6.3.6 Verilog Descriptions of SUM and Short Memory Blocks

The block SUM is in charge of performing the summation of products between constants ω_j and the result of the previous iterations of the state variable x. Figure 6.7 shows the internal hardware for this block. As mentioned in Chap. 2, the approximation of a solution of a fractional-order memristive system must consider the application of the short memory principle, which allows the FPGA implementation. In this manner, the block SUM includes a short memory (Short_Memory_RBM) that is implemented by using PIPO registers. The block also includes a ROM memory to store the constants ω_j; includes a FSM (FSM_XMJ), two type-D flip-flops, a descending counter, and a counter with stop pin; and includes a multiplier-accumulator block, which is described in Chap. 3 and which is the main piece to perform the summation of products between ω_j and x_{m-j}.

The block diagrams and the Verilog descriptions of the blocks ROM, FFD, and MAC have been detailed in Chap. 3. The block MAC2 that is embedded into the block SUM in Fig. 6.7 has the Verilog description quite similar to the one shown in Listing 3.16, with the difference that the block Multiplier_enb is now changed by a combinational multiplier described as block Multiplier_com. In the case of the block ROM, it stores the 127 coefficients ω_j, all of them in fixed-point format.

The block SUM embeds counters, as shown in Fig. 6.7; they behave as a counter with stop signal (Counter_STP) and as a descending counter (Counter_Down). The first one, Counter_STP, is in charge of activating a signal in the state machine

(FSM_XWJ), to know the time when the short memory has filled its 127 cells. The block Counter_Down is in charge as an address pointer for the short memory (Short_Memory_RBM) and also for the block ROM (ROM_WJ). This counter starts counting at 126 and goes down to finish at 0. Once the descending count reaches 0, the block Counter_Down activates a signal to indicate that all summation of products have been performed within the block MAC. The block and Verilog descriptions for block Counter_STP are described in the previous section. The internal hardware and Verilog description of the block Counter_Down are shown in Fig. 6.8 and Listing 6.7.

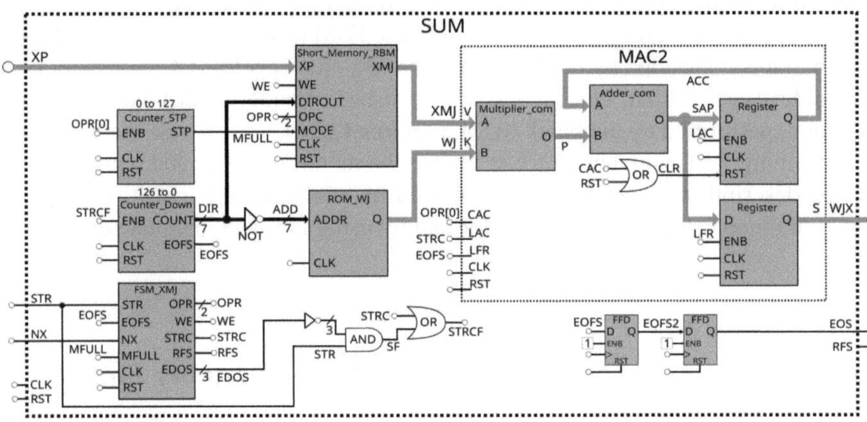

Fig. 6.7 Internal hardware description of the block SUM

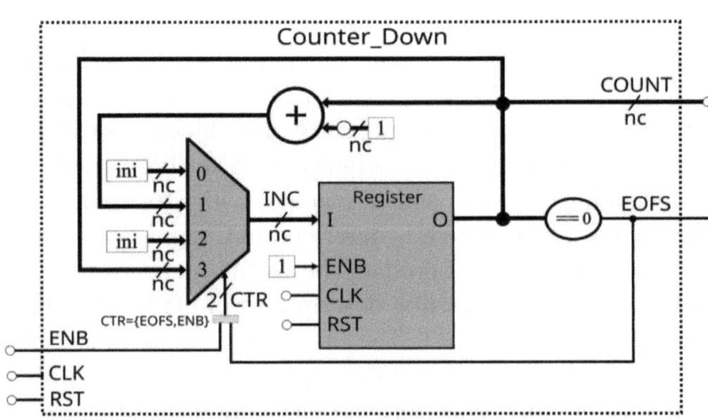

Fig. 6.8 Internal hardware description of the block Counter_Down

Listing 6.7 Verilog description of the block Counter_Down

```
/*-----------------------------------------------------------
---Project: down counter                                  ---
---Inputs: RST - sets the counter to zero                 ---
           CLK - clock signal                             ---
           ENB - enable counter                           ---
---Output: COUNT - gives the count value                  ---
           EOFS - gives '1' when count equals 0           ---
-----------------------------------------------------------*/
module Counter_Down #(parameter nc=7, parameter ini=7'd126)
    (RST,CLK,ENB,COUNT,EOFS);
    input RST,CLK,ENB;
    output [nc-1:0] COUNT;
    output EOFS;

    wire [nc-1:0] DEC;
    wire [1:0] CTR;

    assign EOFS = (COUNT=={nc{1'b0}}) ? 1'b1: 1'b0;

    assign CTR = {EOFS,ENB};
    assign DEC = (CTR==2'd0) ? ini:
                 (CTR==2'd1) ? COUNT - 1'b1:
                 (CTR==2'd2) ? ini:
                               COUNT;

    Register #(nc,ini) U01 (RST,CLK,1'b1,DEC,COUNT);

endmodule
```

The block SUM shown in Fig. 6.7 includes the block labeled as Short_Memory_RBM, which is a short memory for temporal store of data and consists of a counter and a register-based memory (RBM) of 127 cells with 40-bit length. The internal hardware of this memory is described in Fig. 6.9, where the control pines are sketched as external ones in the block with dotted lines.

A detailed description for the design of an RBM 7×32 is given in Chap. 3. In this case study, for the synthesis of the fractional-order memristive system with time delay, an RBM of 127×40 is required. The internal hardware description of this memory is then similar to the one of the RBM shown in Fig. 3.15. The difference is that in the diagram, a decoder block is necessary of 7–128, and it also requires a multiplexer with 128 inputs, a PIPO block with 127 registers, and combinational logic based on gates and multiplexers 2–1. The Verilog description of the block RBM follows the same format given in Listing 3.15, but now, one must include more lines to describe all wires, the instantiation of registers, the assignation of combinational logic, and the assignation of pines in the multiplexer.

The advantage of using an RBM is the capability of programming two store modes. During the first 127 iterations, each new value of the state variable x is stored in a single register, beginning with the first one 1 and ending at 127. The block Counter_Load is responsible of incrementing the memory address to point to

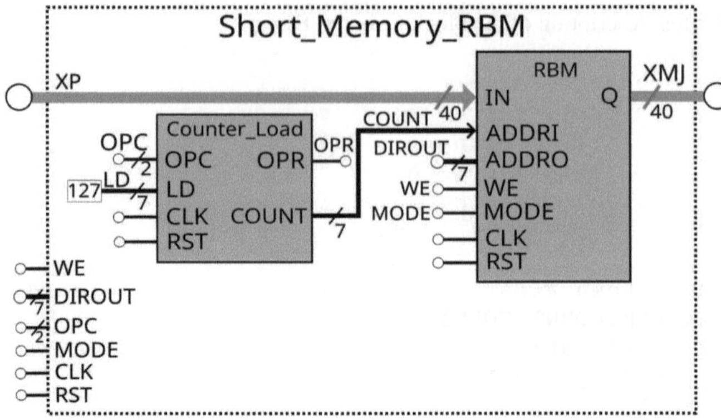

Fig. 6.9 Internal diagram of the block Short_Memory_RBM consisting of a counter and an RBM 127×40)

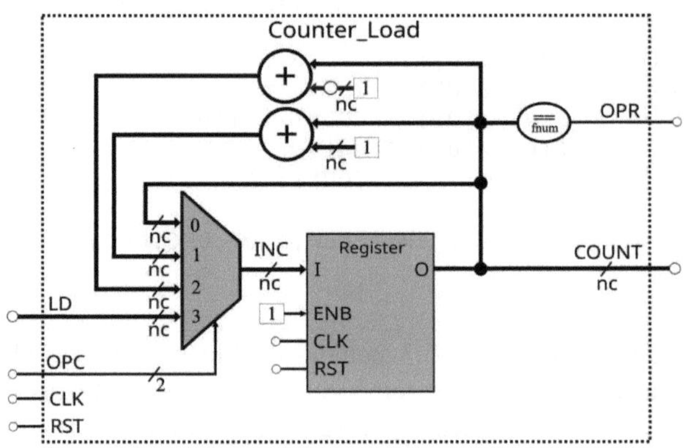

Fig. 6.10 Internal hardware description of block Counter_Load, which is similar to the one for Counter_opc

a single register. When the memory is totally filled, the signal MODE is set to a high state value, thus allowing that the register 127 stores the new value of x, and all the previous results are shifted to the adjacent register. The signal MODE comes from the block Counter_STP (embedded into the block SUM). The block Counter_Load is only in charge to establish the address to the block RBM for the first 127 iterations. When the block RBM is totally filled, the input OPC of the counter allows that the value of the counter remain to 127. The internal hardware description of the block Counter_Load is shown in Fig. 6.10. The Verilog description is similar to the one performed for the block Counter_OPC and given in Listing 3.11, but now the input LD is added, and the order of the operations in the multiplexer changes. The Verilog description of the block Short_Memory_RBM is given in Listing 6.8.

Listing 6.8 Verilog description of the block Short_Memory_RBM

```
/*--------------------------------------------------------------
---Project: Short memory based on RBM                        ---
---Inputs: RST - sets all registers to zero                  ---
           CLK - clock signal                                ---
           WE  - write enable                                ---
           MODE - control memory functionality               ---
           OPC - control counter mode                        ---
           XP  -  data to store in memory                    ---
           DIROUT - memory address to get a value            ---
---Output: XMJ - value from a register                       ---
--------------------------------------------------------------*/
module Short_Memory_RBM #(parameter n=40)
   (RST,CLK,WE,MODE,OPC,XP,DIROUT,XMJ);
   input RST,CLK,WE,MODE;
   input [1:0] OPC;
   input [n-1:0] XP;
   input [6:0] DIROUT;
   output [n-1:0] XMJ;

   wire OPR;
   wire [6:0] COUNT;
   wire [6:0] LD;
   assign LD = 7'd127;

   Counter_Load #(7,127) U1 (RST,CLK,OPC,LD,COUNT,OPR);
               RBM #(n) U2 (RST,CLK,WE,MODE,XP,COUNT,DIROUT,XMJ);

endmodule
```

The block SUM also includes a finite-state machine labeled as FSM_XMJ. It is in charge of controlling the sum of products performed by the block MAC and in charge of updating the block RBM. The block SUM has inputs assigned to the pines STR, EOFS, NX, MFULL (memory full), CLK, and RST. The outputs of SUM are labeled as OPR (operation), WE (write enable), STRC (start down counter), and RFS. The FSM of this block SUM is shown in Fig. 6.11, where one can appreciate the diagram of the transition of states.

The states in Fig. 6.11 perform the following actions to the block SUM:

1. State S0: At this state, the FSM waits for the signal STR to start the summation of products.
2. State S1: The FSM activates the signal STRC to start the counter that points to all address of both memories ROM and RBM. This state ends when all summation of products are performed.
3. State S2: The FSM waits for the signal NX to update the short memory.
4. State S3: The FSM activates the signal WE to store a new datum in the short memory.
5. State S4: The FSM sends the signal OPR to increase the counting in a unity in block Counter_Load. This signal OPR also erases the register accumulator of block MAC2 and increases a unity in the block Counter_STP (only if RBM is not totally filled). After one clock cycle, this block goes to state S0.

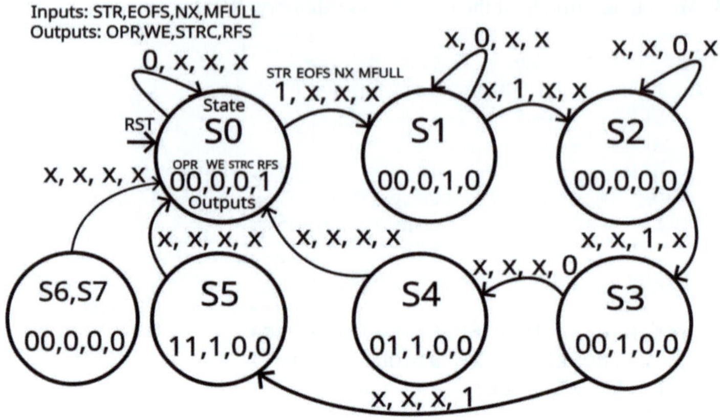

Fig. 6.11 Diagram of transition of the states of the block FSM_XMJ

6. State S5: The FSM sends the signal OPR to hold the count in block Counter_Load to 127. This signal OPR also erases the register accumulator of the block MAC2. After one clock cycle, this block goes to state S0.
7. States S6 and S7: These states return the block FSM to state S0.

The Verilog description of this FSM is based on the template provided in Chap. 3.

Once every internal elements in block SUM are identified, the Verilog description can be performed. The code is given in Listing 6.9, which is based on the internal hardware description shown in Fig. 6.7.

Listing 6.9 Verilog description of the block SUM

```
/*-----------------------------------------------------------------
---Project: SUM block                                          ---
---Inputs: RST - sets all registers to zero                    ---
           CLK - clock signal                                  ---
           STR - start FSM                                     ---
           NX  - next X value                                  ---
           XP  - data to store in memory                       ---
---Output: WJX - value of SUM                                  ---
           EOS - end of SUM                                    ---
           RFS - ready for SUM                                 ---
-----------------------------------------------------------------*/
module SUM #(parameter n=40,parameter frac=34)
    (RST,CLK,STR,NX,XP,WJX,EOS,RFS);
    localparam nc=7;
    localparam ini=126;     //Accumulative sums.
    localparam fnum=127;    //Memory limit.

    input RST,CLK,STR,NX;
    input [n-1:0] XP;
    output [n-1:0] WJX;
    output EOS,RFS;

```

```
    wire [1:0] OPR;
    wire WE,STRC,EOFS,EOFS2,MFULL;
    wire [nc-1:0] DIR,ADD;
    wire [2:0] EDOS;
    wire SF,STRCF;
    wire [n-1:0] WJ;
    wire [n-1:0] XMJ;

    assign SF = STR & ~EDOS[0] & ~EDOS[1] & ~EDOS[2];
    assign STRCF = STRC | SF;
    assign ADD = (~DIR);

    FSM_XMJ U01 (RST,CLK,STR,EOFS,NX,MFULL,OPR,WE,STRC,RFS,EDOS);
    Counter_Down #(nc,ini) U02 (RST,CLK,STRCF,DIR,EOFS);
    Counter_STP #(nc,fnum) U03 (RST,CLK,OPR[0],MFULL);
          Short_Memory_RBM U04 (RST,CLK,WE,MFULL,OPR,XP,DIR,XMJ);
            ROM_WJ #(n,nc) U05 (CLK,ADD,WJ);
    MAC2 #(n,frac) U06 (RST,CLK,STRC,EOFS,OPR[0],XMJ,WJ,WJX);
               FFD U07 (RST,CLK,1'b1,EOFS,EOFS2);
               FFD U08 (RST,CLK,1'b1,EOFS2,EOS);

endmodule
```

6.3.7 *Verilog Description of Block FRAC_MEM_TD_EQU*

The Verilog description shown in this section uses the previous blocks and the basic ones given in Chap. 3. In this case, a higher hierarchy is described for the block labeled as FRAC_MEM_TD_EQU. The Verilog description given in Listing 6.10 is based on the internal hardware description shown in Fig. 6.1. This description is just for one iteration; in the next section, the complete iterative process is described.

Listing 6.10 Verilog description of the block FRAC_MEM_TD_EQU

```
/*---------------------------------------------------------------
---Project: Fractional Memristive Time-Delay                 ---
   Fixed-point format: 1:5:34                                ---
---Inputs: RST - sets the output with the initial condition  ---
           CLK - clock signal                                ---
            ST - start new iteration                         ---
---Output: RX,RXT - logic output of n bits                   ---
-------------------------------------------------------------*/
module FRAC_MEM_TD_EQU (RST,CLK,STF,X0,XP,RX,RXT,LR);
   localparam n=40;
   localparam frac=34;

   input RST,CLK,STF;
   input [n-1:0] XP,X0;
   output [n-1:0] RX, RXT;
   output LR;

```

```
    wire [n-1:0] WJXMJ, PSI, S1, RS, XP, XT, FXT;
    wire EOS, RFS, STR, NX, SEL;

                 FSM_FR U01 (RST,CLK,STF,EOS,RFS,STR,LR,NX);
          SUM #(n,frac) U02 (RST,CLK,STR,NX,XP,WJXMJ,EOS,RFS);
    Counter_STP #(9,350) U03 (RST,CLK,STR,SEL);
         Memory_XT #(n) U04 (RST,CLK,LR,XP,XT);

      FUNC_FXT #(n,frac) U05 (RST,CLK,SEL,XT,FXT);
          FUNC_PSI #(n) U06 (RST,CLK,XP,FXT,PSI);
       FUNC_S1 #(n,frac) U07 (RST,CLK,PSI,X0,S1);

         Subtractor #(n) U08 (RST,CLK,S1,WJXMJ,RS);

         Register #(n,0) U09 (RST,CLK,LR,RS,RX);
         Register #(n,0) U10 (RST,CLK,LR,XT,RXT);

endmodule
```

6.4 Synthesis of a PRNG Based on a Fractional-Order Memristive System with Time Delay

The Verilog description given in the previous section does not have the feedback loop to perform iterations. Therefore, it is necessary to connect the output of the block SUM to its inputs by adding a control stage and to synthesize a PRNG based on the times series of the fractional-order memristive system with time delay.

This section describes the simulated and experimental results of a PRNG based on the fractional-order 1D memristive time-delay chaotic system. The mathematical model of the memristive system is simulated within Matlab to observe the chaotic time series and the attractor. The Verilog description is detailed in the previous section, and it is simulated in the program Active-HDL through the testbench. This section describes the synthesis of a PRNG based on the fractional-order memristive system and provides the hardware resource consumption by synthesizing the PRNG into an FPGA from Altera/Intel.

6.4.1 Time Series and Attractors Simulated within Matlab

The simulation of the fractional-order 1D memristive system with time delay can be performed by using Matlab. Recall that a simulation is necessary to determine the maximum and minimum values of the amplitudes of the signals to establish the fixed-point format and afterwards to design the digital blocks. Matlab provides a computer arithmetic of a floating point with 64 bits. In addition, for the FPGA implementation, fractional-order chaotic systems must be simulated by applying

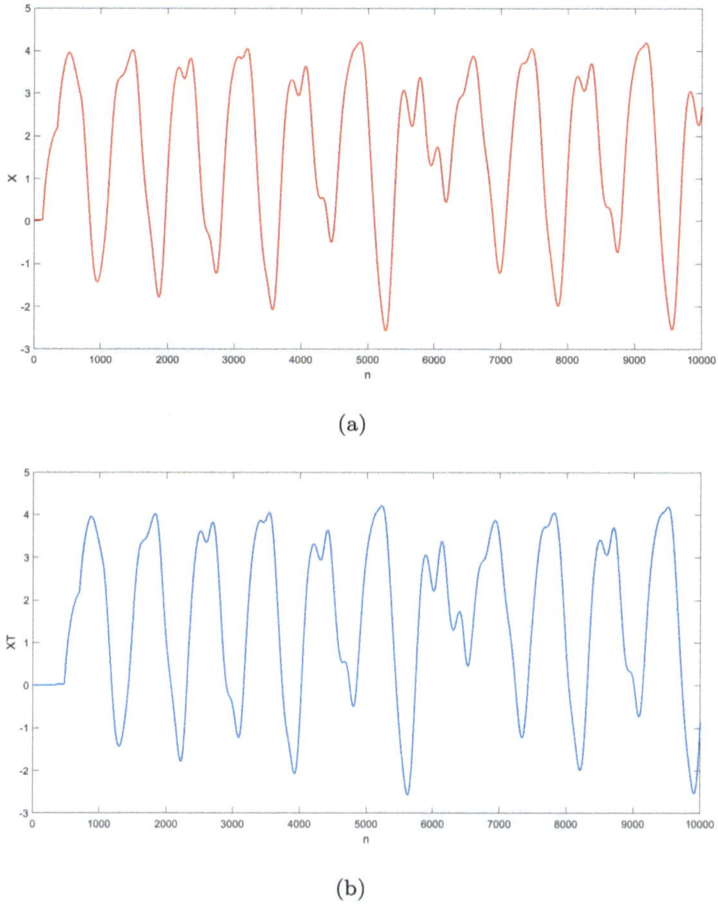

Fig. 6.12 Simulation of the fractional-order memristive system with time delay, generating the time series for the variables (**a**) x_n and (**b**) $x_{\tau n}$

the short memory principle described in Chap. 2. The determination of the length of memory is important to synthesize the summation of products, as detailed in the previous sections.

The simulation of the fractional-order memristive system provides the chaotic time series shown in Fig. 6.12, for the state variable x_n and the delayed one $x_{\tau n}$. The simulation was performed considering 10,000 iterations by setting the initial condition to $x_0 = 0$. Figure 6.13 shows two types of attractors that are generated by varying parameter b in Eq. (6.2), while the other parameters remain in their values by now performing 100,000 iterations.

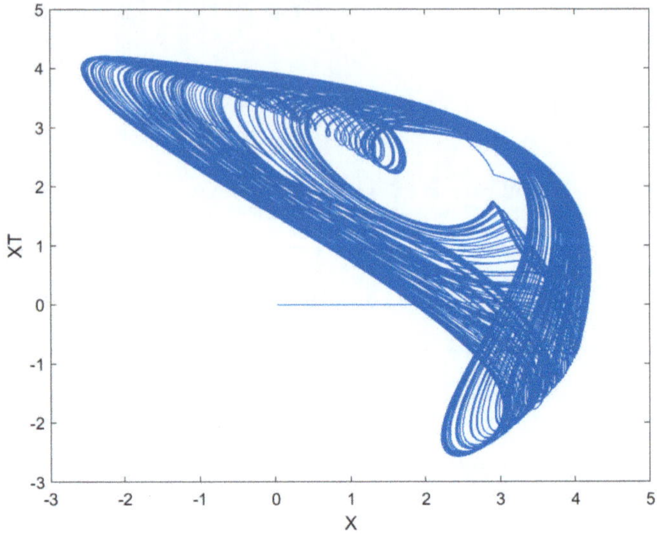

(a)

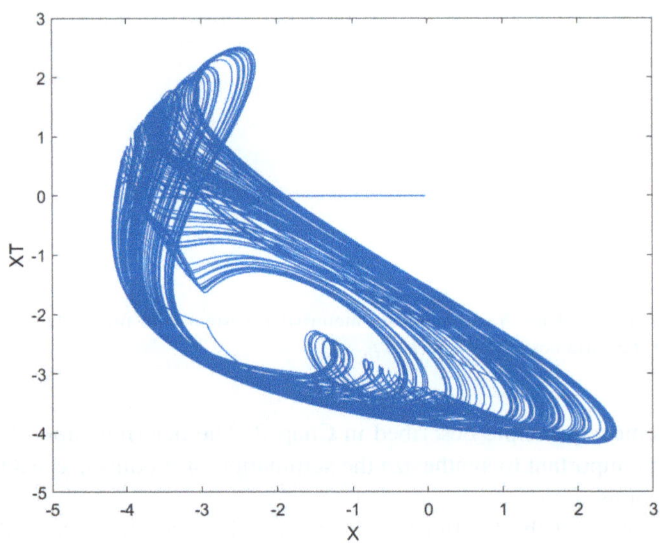

(b)

Fig. 6.13 Chaotic attractors simulated by setting $a = 1.0$, $\delta = 0.1$, $\gamma = 1.0$, $\mu = 0.98$, $h = 0.01$, and $m = 128$, for two different values of parameter b: (**a**) $b = 4.5$ and (**b**) $b = -4.5$

6.4.2 PRNG from Chaotic Time Series by Applying mod256

It is possible to generate chaotic time series in binary format by taking the eight least significative bits (LSBs) of each value at each iteration when it is represented in fixed-point format. This leads us to synthesize a PRNG by concatenating all eight LSBs of each datum in the chaotic time series. However, one must perform NIST and TestU01 tests, to evaluate the randomness, as described in Chap. 4. The randomness can be improved by adding a post-processing stage. In this manner, this section applies the operation with mod256 to increase the randomness and then to guarantee the synthesis of a good PRNG. The post-processing stage performs the following operations: The operation mod256 is applied to each real value in the chaotic time series and the eight LSBs are taken and concatenated to generate a chaotic binary string. One may perform a conversion from floating point to fixed-point format.

6.4.3 Verilog Description of a PRNG and Simulation Results

The Verilog description of a PRNG based on chaotic time series requires to design the hardware to connect the outputs of the block diagram of the fractional-order 1D memristive system with time delay shown in Fig. 6.1. Basically, the outputs must be fed back to the inputs by designing an appropriate control stage. For instance, the Verilog description given in Listing 6.10 is based on the internal hardware description shown in Fig. 6.1. This description is just for one iteration, and the complete iterative process is described in the following subsections.

6.4.3.1 Block Description

The architecture that describes the fractional-order memristive system shown in Fig. 6.1 requires of a certain number of clock cycles (CCs) to generate a datum. The feedback of the outputs to the inputs in that figure can generate the chaotic oscillations. In this case, the feedback must connect the output (RX) to the input of the block, labeled as (XP). It is also necessary to design the block circuitry to assign the initial condition (X0) and to activate the start signal (STF). The block description of this process is shown in Fig. 6.14, where the PIPO register processes the initial condition X0 that is sent to the main block which includes a feedback to generate the binary sequences that have chaotic behavior.

The iterations are then executed by the block description shown in Fig. 6.14, where the PIPO register processes the initial condition just for the first iteration, and from the second iteration, the bus XOF is disconnected, so that the block FRAC_MEM_TD_OSCILLATOR remains iterating with an appropriate number of

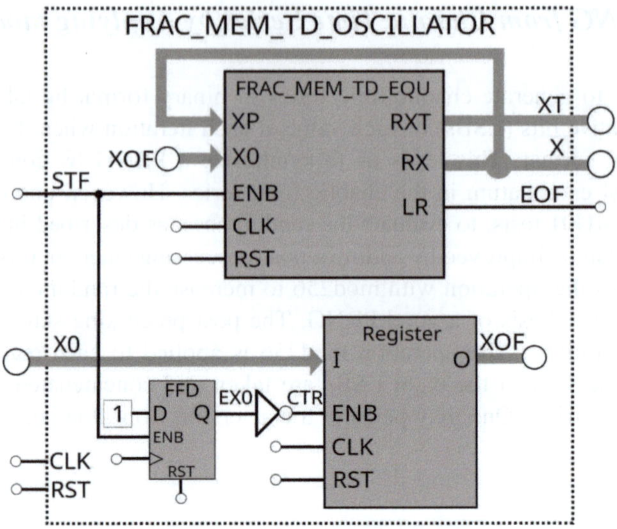

Fig. 6.14 Block description of the fractional-order 1D memristive system with time delay based on the block FRAC_MEM_TD_OSCILLATOR

CCs to save every valid data in the state variables x_n labeled as X and $x_{\tau n}$ labeled as XT. To change the initial condition in Fig. 6.14, a signal is applied to reset (RST) the register. The Verilog description of the PRNG based on the fractional-order memristive system with time delay shown in Fig. 6.14 is given in Listing 6.11.

Listing 6.11 Verilog description of the block FRAC_MEM_TD_OSCILLATOR

```
/*----------------------------------------------------------
---Project: Complete Fractional Memristive Time-Delay     ---
---Fixed-point format: 1:5:34                            ---
---Inputs:  RST - sets the outputs to zero               ---
            CLK - clock signal                           ---
            STF - starts a new iteration                 ---
            X0  - initial condition                      ---
---Outputs: X,XT- output state variables                 ---
           EOF: end of function                          ---
----------------------------------------------------------*/
module FRAC_MEM_TD_OSCILLATOR (RST,CLK,STF,X0,X,XT,EOF);
    localparam n=40;
    input RST,CLK,STF;
    input [n-1:0] X0;
    output [n-1:0] X,XT;
    output EOF;

    wire [n-1:0] RX,RXT,XOF;
    wire OPR,CTR,EX0;

    assign CTR = ~EX0;

```

```
    FRAC_MEM_TD_EQU U01 (RST,CLK,STF,X0F,X,X,XT,EOF);
    Register #(n,0) U02 (RST,CLK,CTR,X0,X0F);
                FFD U03 (RST,CLK,STF,1'b1,EX0);

endmodule
```

6.4.3.2 Verilog Simulation

The Verilog description of the fractional-order memristive system with time delay can be simulated under the environment Active-HDL by performing a testbench. This environment includes a waveform tool, which allows to observe the behavior of the signals. In fact, this software also allows to perform simulations considering computer arithmetic of fixed-point format. This simulation process on Active-HDL is important to verify the correct function of the complete hardware description of the fractional-order memristive system with time delay. The Verilog code to perform the testbench is given in Listing 6.12. The simulated time series under Active-HDL are shown in Fig. 6.15.

Listing 6.12 Verilog description of the testbench for the block FRAC_MEM_TD_OSCILLATOR

```
/*---------------------------------------------------------------
---Project: Testbench for Fractional Memristive Time-Delay    ---
---Fixed-point format: 1:5:34                                 ---
---Inputs:  RST - sets the outputs to zero                    ---
            CLK - clock signal                                ---
            STF - starts a new iteration                      ---
            X0  - initial condition                           ---
---Outputs: X,XT- output state variables                      ---
           EOF: end of function                               ---
---------------------------------------------------------------*/
'timescale 1 ns/1 ps
'define ciclos 10
module TB_FRAC_MEM_TD_OSCILLATOR();
    parameter ITER = 16'd10000;
    parameter n=40;
    reg RST,CLK,ST;
    reg [n-1:0] X0;
    wire [n-1:0] XI, XT;
    integer file, file1;

   FRAC_MEM_TD_OSCILLATOR U1 (RST,CLK,ST,X0,XI,XT,EOF);
   always #'ciclos CLK = ~CLK;

   initial
    begin
        CLK=0; RST=1; ST=1; #0; RST=0;
        X0=40'd0;
        file  = $fopen("FR_MEM_RX_F.txt");
        file1 = $fopen("FR_MEM_RXT_F.txt");
        #10;
```

```

      for (integer i = 0; i < ITER; i = i + 1)
        begin
         $fwrite(file,  "%h\n", XI);
         $fwrite(file1, "%h\n", XT);
         #2700;     // 135 CLK
        end

     $finish;
     $fclose(file);
     $fclose(file1);
   end
endmodule
```

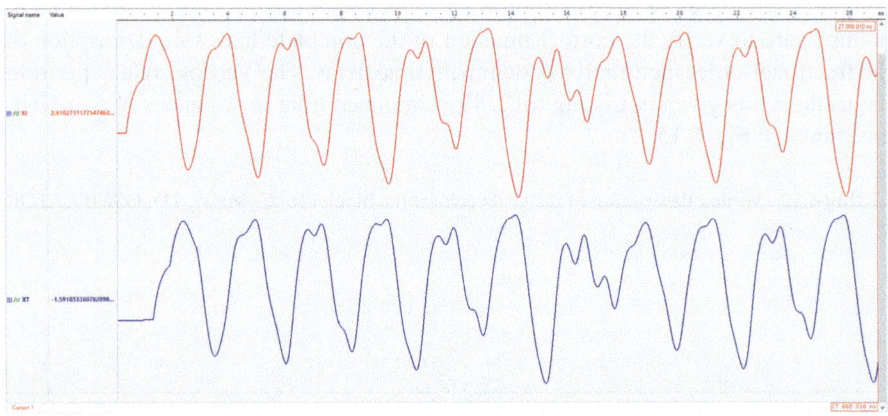

Fig. 6.15 Chaotic time series provided from the simulation of the fractional-order 1D memristive system under Active-HDL

The testbench provides two text files *.txt with the results in hexadecimal format for each iteration to update the state variables x_n labeled as X and $x_{\tau n}$ labeled as XT. The text files can be read within Matlab to observe the chaotic time series and the attractors generated from the digital hardware description.

6.5 PRNG Based on Fractional-Order Chaotic System: FPGA Implementation and Experimental Results

The FPGA synthesis of the fractional-order memristive system shown in Fig. 6.14 requires the design of additional digital blocks to allow the observation of the experimental data on an oscilloscope. If one observes the time series before their digitalization (binary strings) and the chaotic attractors, a digital-to-analog (DAC) converter is required, which in this book has 16 bits. The connection of the outputs

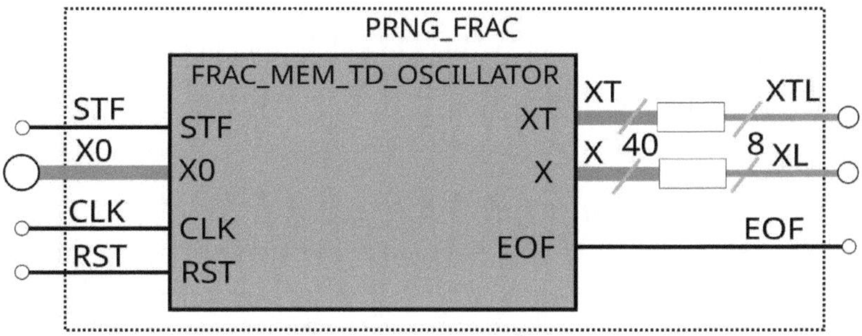

Fig. 6.16 PRNG based on the fractional-order memristive system with time delay

Table 6.2 FPGA Cyclone IV EP4CGX150DF31C7 hardware resource consumption

Resources	Used	Available
Logic elements (LE)	9,878	149,760
Combinational functions	9,769	149,760
Dedicated logic registers	5,578	149,760
Memory bits	19,000	6,635,520
Embedded multiplier 9-bit	57	720
Maximum frequency (MHz)	56.9	–
Latency at 50 MHz (ns)	2700	–
Throughput at 50 MHz (Mbps)	5.92	–
Efficiency (Kbps/LE)	0.599	–

of the FPGA with the DAC requires the connection of a communication block with the protocol like the serial peripheral interface (SPI).

To implement a PRNG into the FPGA, one can concatenate the eight LSBs of each datum generated at each iteration in the fractional-order memristive system. Figure 6.16 shows the PRNG based on the fractional-order memristive system. The state variables x_n labeled as X and $x_{\tau n}$ labeled as XT have length of 40 bits in fixed-point format. These buses are truncated to the eight LSBs to produce the outputs labeled as XL and XTL, who generates the random binary strings.

In this book, the block PRNG_FRAC is synthesized by using an FPGA Cyclone IV, whose hardware resource consumption is summarized in Table 6.2.

6.5.1 Experimental Time Series and Chaotic Attractor

Figure 6.17 shows the experimental time series observed in a Teledyne Lecroy oscilloscope.

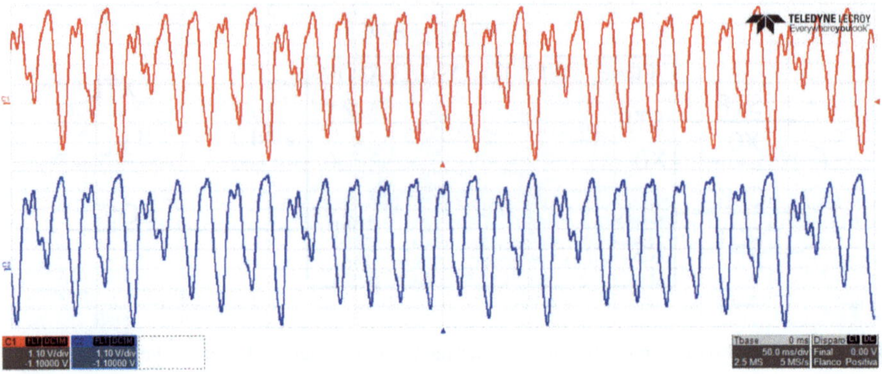

Fig. 6.17 Experimental chaotic time series: x_n in red color and $x_{\tau n}$ in blue color

Figure 6.18 shows the experimental chaotic attractors observed in a Teledyne Lecroy oscilloscope.

6.5.2 Experimental Generation of Binary Strings

The behavior of the PRNG shows the binary strings for each pin associated to the eight LSBs taken from each datum generated at each iteration to obtain x_n labeled as X and $x_{\tau n}$ labeled as XT. Figure 6.19 shows the experimental behavior of the very LSB observed in a Teledyne Lecroy oscilloscope. This binary string is generated by the block PRNG_FRAC that is synthesized in the FPGA Cyclone IV EP4CGX150DF31C7.

6.6 Conclusion

The conclusion of this chapter is that the FPGA synthesis of a fractional-order chaotic system, as the memristive one with time delay, requires a higher design effort compared to the FPGA synthesis of PRNGs based on integer-order chaotic systems or maps. The generated random binary strings can be applied to develop applications like image encryption and authentication, as shown in Chap. 7.

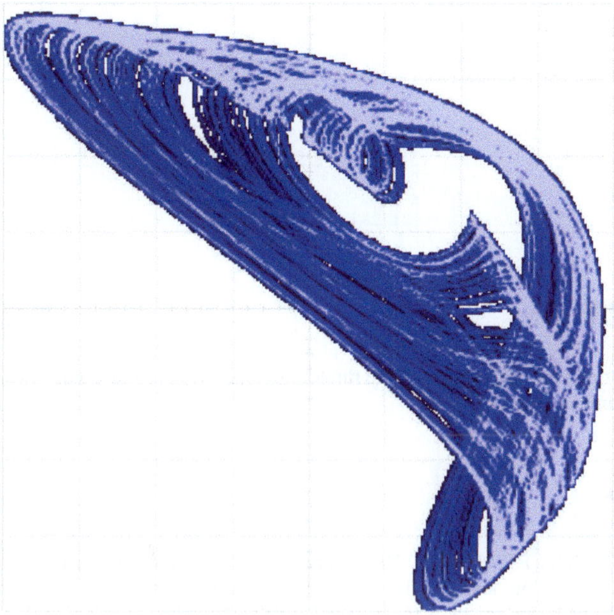

(a)

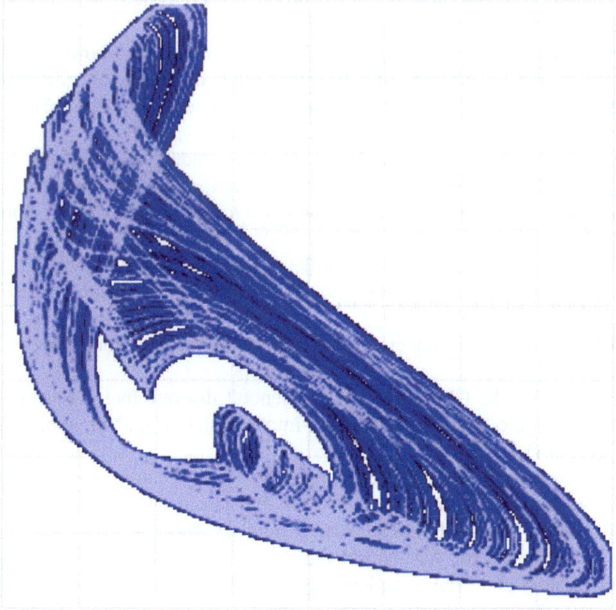

(b)

Fig. 6.18 Experimental attractors $x_n - x_{\tau n}$ generated with the parameters set to $a = 1.0$, $\delta = 0.1$, $\gamma = 1.0$, $\mu = 0.98$, $h = 0.01$, and $m = 128$, for two values of b: (**a**) $b = 4.5$ and (**b**) $b = -4.5$

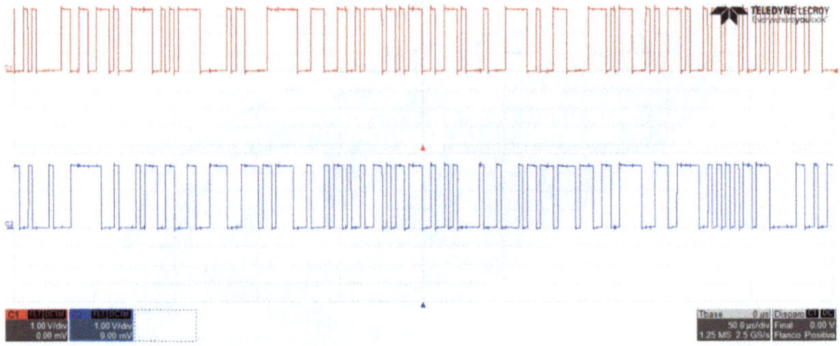

Fig. 6.19 Experimental observation of the random binary strings for the state variables x_n (in red color) and $x_{\tau n}$ (in blue color)

References

1. J. Hwang, G. Kale, P.P. Patel, R. Vishwakarma, M. Aliasgari, A. Hedayatipour, A. Rezaei, H. Sayadi, Machine learning in chaos-based encryption: theory, implementations, and applications. IEEE Access **11**, 125749–125767 (2023)
2. O.I. Tacha, J.M. Munoz-Pacheco, E. Zambrano-Serrano, I.N. Stouboulos, V.-T. Pham, Determining the chaotic behavior in a fractional-order finance system with negative parameters. Nonlinear Dyn. **94**(2), 1303–1317 (2018)
3. A.D. Pano-Azucena, B. Ovilla-Martinez, E. Tlelo-Cuautle, J.M. Muñoz-Pacheco, L.G. de la Fraga, Fpga-based implementation of different families of fractional-order chaotic oscillators applying grünwald–letnikov method. Commun. Nonlinear Sci. Num. Simul. **72**, 516–527 (2019)
4. E. Tlelo-Cuautle, A. Pano-Azucena, O. Guillén-Fernández, A.S. Juárez, *Analog/Digital Implementation of Fractional Order Chaotic Circuits and Applications* (Springer, Berlin, 2020)
5. I. Petráš, The fractional-order lorenz-type systems: a review. Fract. Calcul. Appl. Analy. **25**(2), 362–377 (2022)
6. E. Tlelo-Cuautle, L.G. De La Fraga, O. Guillén-Fernández, A. Silva-Juárez, *Optimization of Integer/Fractional Order Chaotic Systems by Metaheuristics and Their Electronic Realization* (CRC Press, Boca Raton, 2021)
7. L.O. Chua, Memristor the missing circuit element. IEEE Trans. Circuit Theory **18**(5), 507–519 (1971)
8. B. Xu, S. Zou, L. Bai, K. Chen, J. Zhao, A general discrete memristor emulator based on taylor expansion for the reconfigurable fpga implementation and its application. Nonlinear Dyn. **2023**, 1395–1414 (2023)
9. Z. Kotadai, C. Fischer, J. Rodrí-guez-Muñoz, E. Tlelo-Cuautle, E. Yves, Multistability and initial-offset boosting dynamics in a new 3D memristive chaotic system with FPGA implementation. Eur. Phys. J. Plus **139**, Article number 70 (2024)
10. X. Zhang, G. Yang, S. Liu, A.J. Moshayedi, Fractional-order circuit design with hybrid controlled memristors and FPGA implementation. AEU Int. J. Electr. Commun. **153**, 154268 (2022)
11. J. Zhang, X. Wang, J. Hou, Y. Guo, Q. Xie, A novel memristor-based multi-vortex hyperchaotic circuit design and its application in image encryption. J. Circuits Syst. Comput. **33**, 2450100 (2023)

12. A.L.M. Biamou, V.K. Tamba, F.K. Tagne, A.C.N. Takougang, Fractional-order-induced symmetric multi-scroll chaotic attractors and double bubble bifurcations in a memristive coupled hopfield neural networks. Chaos Solit. Fract. **178**, 114267 (2024)
13. V.-T. Pham, A. Velichko, V.V. Huynh, A.V. Radogna, G. Grassi, S.M. Boulaaras, S. Momani, Analysis of memristive maps with asymmetry. Integr. VLSI J. **94**, 102110 (2024)
14. X. Kong, F. Yu, W. Yao, S. Cai, J. Zhang, H. Lin, Memristor-induced hyperchaos, multiscroll and extreme multistability in fractional-order hnn: Image encryption and fpga implementation. Neural Netw. **171**, 85–103 (2024)
15. M. Kumar, An efficient numerical scheme for solving a fractional-order system of delay differential equations. Int. J. Appl. Comput. Math. **8**(5), 1–18 (2022)
16. Z. Kotadai, E.J. Yves, C. Fischer, J.D. Rodríguez-Muñoz, M.F. Moreno-Lopez, E. Tlelo-Cuautle, J. De Dieu Nkapkop, Fractional order 1D memristive time-delay chaotic system with application to image encryption and FPGA implementation. Math. Comput. Simul. **227**, 58–84 (2024). https://doi.org/10.1016/j.matcom.2024.07.035

Chapter 7
Chaos-Based Encryption and Authentication of RGB Images

7.1 Introduction

Chapter 1 of this book described the basics of chaos-based secure communications and some software codes to generate random binary sequences or numbers. A brief description of hardware implementations of PRNGs based on chaotic maps and integer- and fractional-order chaotic systems was also given. In particular, one can find details of implementations in [1] and in [2], where one can also find optimization algorithms to enhance randomness of chaotic systems. It was also mentioned that some research is currently performed by combining chaotic systems and machine learning algorithms, as shown in [3].

Chapter 2 showed the application of numerical methods to discretize the mathematical models of continuous-time chaotic systems of integer- and fractional-order. Some mathematical models for 3D, 4D, and 5D chaotic systems were presented, as well as for a memristor-based fractional system with time delay, for which a more elaborated numerical method was described. As the goal of this book is the design of PRNGs based on chaos, one way to improve randomness is by evaluating the Lyapunov exponents and Kaplan-Yorke dimension, as shown in [1, 2]. In this book, the randomness is evaluated by performing three statistical tests known as NIST, TestU01, and Dieharder, as given in Chap. 4. However, it was mentioned that Dieharder tests can be considered within the most used ones, the NIST and TestU01.

Once a mathematical model describing a chaotic system is evaluated to generate randomness, this book detailed the hardware description and simulation under Verilog code, as detailed in Chap. 3. The Verilog codes of basic arithmetic blocks, such as adders, subtractors, and multipliers, and of sequential blocks were used in high-level descriptions by instantiation processes. In this manner, the design of PRNGs based on chaotic maps and systems was detailed in Chap. 5, and the design based on fractional-order chaotic systems was given in Chap. 6. Both chapters also showed the FPGA synthesis and experimental observation of chaotic time series, strange attractors, and the generation of binary strings. Now, in this chapter, the

L. G. de la Fraga et al., *Random Number Generators*,
https://doi.org/10.1007/978-3-031-82865-2_7

application of those chaos-based PRNGs is shown by authenticating encrypted RGB images. It is important to mention that at the time of producing this book, the Verilog descriptions can be implemented on an integrated circuit design project, e.g., by considering the open-source code projects as the one offered by Tiny Tapeout https://tinytapeout.com.

7.2 Secure Communication System Using Chaos-Based PRNGs

The first chaos-based secure communication system was possible to implement after the seminal work given in [4], which was introduced for the synchronization of chaotic systems and which is the most cited work of Pecora and Carroll, in physical review letters. Variants and improvements of this master-slave or transmission-reception topology have been developed, so that the synchronized systems can be applied in a higher level to authenticate encrypted RGB images, as shown in [5], where chaotic binary strings are combined with hash functions.

The system for the authentication of encrypted data described in this chapter can be programmed on microcontrollers and other embedded devices for wireless applications, Internet of Things, and so on. In fact, it is quite natural that one can have more than one electronic device connected to a wireless communication network, where a huge number of devices can have access to the transmitted data. Some users, labeled as hackers, can have a device with the capability of accessing sensible data from a public channel, so that the information can be available to any users of the wireless network. This imposes the challenge for the development of alternatives to protect and guarantee security of the information that is transmitted on public channels. One solution is the encryption of the information by using chaos-based PRNGs, and in addition, the encrypted data can be authenticated by using hash functions, as shown in this chapter.

The process of encryption and decryption of data can be as complex or as simple, but the goal is that the information transmitted over a public channel can only be recovered if the user has the correct key. In this case, the key is composed of the parameters of the mathematical model of a chaotic map or integer- or fractional-order chaotic system and the seed or initial conditions. Therefore, a chaos-based PRNG can be used to performing XOR operations between the binary strings and the data being transmitted, such as text data, audio, video, or RGB images. The encryption and decryption processes by using chaos-based PRNGs are sketched in Fig. 7.1, where one can appreciate the main blocks labeled as transmitter and receiver. In this chapter, both PRNGs in the transmitter and receiver must be the same. If they are different, then one must apply a synchronization method to control that the slave behaves as the master, as shown in [6], where the secure communication system is implemented on Raspberry Pi devices that are communicated over a WiFi network.

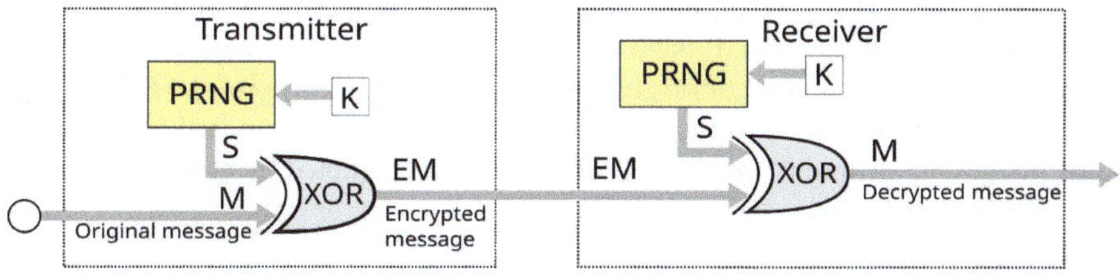

Fig. 7.1 Block diagram description of a secure communication system consisting of chaos-based PRNGs and XOR operations to encrypt and decrypt data

As detailed in Chap. 3, the design of a digital block must be associated to the number of bits to represent the numbers in fixed-point format. In this chapter, the PRNGs are synthesized by blocks having buses of 24 bits. The transmission-reception scheme shown in Fig. 7.1 has the same PRNGs and XOR blocks with buses of 24 bits. In this manner, the PRNG in the transmission stage is activated with the seed for the chaotic map or the initial conditions for the chaotic system, and then it generates the random binary strings. As the encryption and decryption system works with RGB images, one must convert the primary colors, namely, red, green, and blue to pixels that have a byte or eight-bit length. One must calculate the number of total bits for the size of the image, and then, the PRNG must generate such a quantity of bits to perform the XOR operation between each pixel blocks of eight bits generated in the PRNG. This process is performed until the total number of pixels in the RGB image has been transmitted over a public communication channel, as WiFi. As the XOR between the PRNG and RGB image produces a uniform distribution of data, then the receiver block must perform the inverse process very exactly. Thus, in the receiver block, the PRNG is also activated or turned on with the same seed to perform the XOR operation between blocks of eight bits from the transmitter and the PRNG. After decrypting the data taken from the communication channel, the original RGB image can be recovered.

The key point in this type of secure communication topology is how to share the public keys for the activation of the PRNGs and the XOR operation process between the transmitter and receiver blocks. As mentioned above, one way is the application of synchronization methods for chaotic systems, as the ones already described in [1, 6]. The synchronization methods are useful to control the behavior of two systems, but one must also need to identify the number of iterations that must be deleted until the synchronization error is zero.

7.3 Authentication of RGB Images

In addition to the encryption of RGB images, one must guarantee that the transmitted image is not corrupted during the transmission over the public channel. This chapter shows that a hash function can be useful to verify the authenticity of the recovered image. This process basically consists of comparing hash tags or values provided by a given function. If the hash value in the recovered image is the same as the one from the transmitted or original image, then the integrity of the RGB image is not corrupted; otherwise, if the hash values between the transmitted and received RGB images are different, then the RGB image was corrupted by at least one pixel, as shown in this section.

7.3.1 Hash Function Based on the Pseudo Dot Product

The hash function that is used to verify the authenticity of a transmitted RGB image is based on the pseudo dot product (PDP), as already shown in [5], where chaotic binary strings are combined with hash functions. The algorithm to create the hash tag or value basically concatenates the values of seven variables labeled as $(H1, H2, ..., H7)$. In this case, each variable H_i has a length in bits that is equal to the length of bits that produce the chaos-based PRNG. Then each variable H_i provides its value after each iteration that is performed in the XOR operations, multiplications, and modulo operations with a long prime number (n_p). This is performed between the information being authenticated and the random number sequence. Equation (7.1) shows the PDP operation, where letter M represents the messages (in this chapter, the pixels of an image in byte array), and letter S represents a sequence of random numbers:

$$PDP(M, S) = H_1//H_2//H_3//H_4//H_5//H_6//H_7 \tag{7.1}$$

On the other hand, Eq. (7.2) details the operations that must be performed for each variable H_i:

$$\begin{aligned}
H_1 =& [(M_1 \oplus S_1)(M_2 \oplus S_2) \bmod n_p] \oplus [(M_3 \oplus S_3)(M_4 \oplus S_4) \bmod n_p] \oplus \\
& \cdots \oplus [(M_{m-1} \oplus S_{m-1})(M_m \oplus S_m) \bmod n_p], \\
H_2 =& [(M_1 \oplus S_3)(M_2 \oplus S_4) \bmod n_p] \oplus [(M_3 \oplus S_5)(M_4 \oplus S_6) \bmod n_p] \oplus \\
& \cdots \oplus [(M_{m-1} \oplus S_{m+1})(M_m \oplus S_{m+2}) \bmod n_p], \\
H_3 =& [(M_1 \oplus S_5)(M_2 \oplus S_6) \bmod n_p] \oplus [(M_3 \oplus S_7)(M_4 \oplus S_8) \bmod n_p] \oplus \\
& \cdots \oplus [(M_{m-1} \oplus S_{m+3})(M_m \oplus S_{m+4}) \bmod n_p], \\
& \ldots\ldots, \\
H_7 =& [(M_1 \oplus S_{13})(M_2 \oplus S_{14}) \bmod n_p] \oplus [(M_3 \oplus S_{15})(M_4 \oplus S_{16}) \bmod n_p] \oplus \\
& \cdots \oplus [(M_{m-1} \oplus S_{m+11})(M_m \oplus S_{m+12}) \bmod n_p]
\end{aligned} \tag{7.2}$$

Equation (7.2) details that the variables H_i use a different sequence S in the mathematical operations. For this reason, it is necessary to store or memorize the values of 14 iterations in the PRNG, as well as two messages M. For example, during the first calculation of each variable, H_1 performs the product $(M_1 \oplus S_1)$ with $(M_2 \oplus S_2)$, while H_7 performs the product $(M_1 \oplus S_{13})$ with $(M_2 \oplus S_{14})$. Between these two H_i there will always exist a difference of 14 iterations in the generated sequences of random numbers.

The multiplication operation between two variables of n-bit length produces results with almost the double length or almost $2n$ bits, so that one must apply a modulo operation to truncate the unnecessary bits and then to produce the same number of n bits in the results. The algorithm must be executed by using prime numbers to provide better resistance to the collision with the hash values. Although it can also work with number of powers of 2, that is, 2^n. In either way, the chosen number to perform the module operation must truncate the larger number of bits, to the size of the buses, e.g., to n bits.

7.3.2 Operation of the Authentication Stage

In a general sense, the authentication system based on a hash function performs the following operations: The transmission block generates the hash value or tag from the original message M or pixels before the XOR operation (without encryption) and combines them with the random sequences generated by the chaos-based PRNG. After the encrypted message is sent, the transmitter sends the calculated hash value. Afterwards, the reception block decrypts each message and at the same time executes the same algorithm for the hash function with the decrypted data and the same random sequence generated by the PRNG. With these data it generates a new authentication tag. When all the pixels of the encrypted message are received, the receiver gets the hash value provided by the transmitter. Finally, both hash values, the one provided by the transmitter and the one generated in the receiver block, are compared. If both hash tags are the same, the message was not corrupted during the transmission over the public channel. Otherwise, if the hash values are different, or if at least one message block is altered (as one pixel), then both hash tags will be totally different. Figure 7.2 shows the process of encryption, decryption, and authentication of RGB images. It can be appreciated the use of multiplexers to change the route to get the hash value that is coming from the transmitter block.

7.4 Verilog Descriptions of the System for Authentication of Encrypted RGB Images

The block description shown in Fig. 7.2 has the Verilog description given in this chapter. The Verilog code allows to encrypt and authenticate an RGB image. The transmitter block is implemented into an FPGA, which is the responsible to

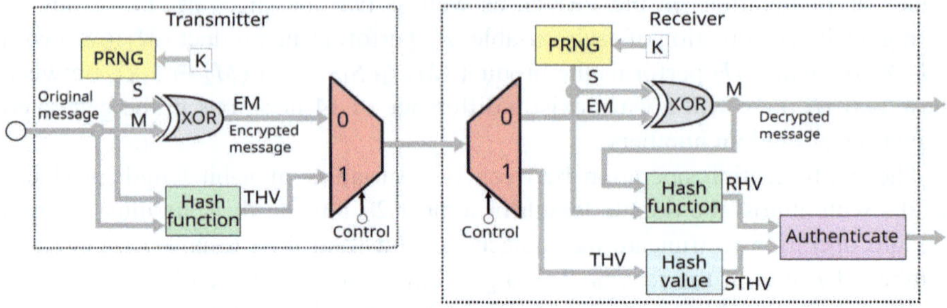

Fig. 7.2 General block description of the encryption and authentication system based on chaos-based PRNGs and hash functions

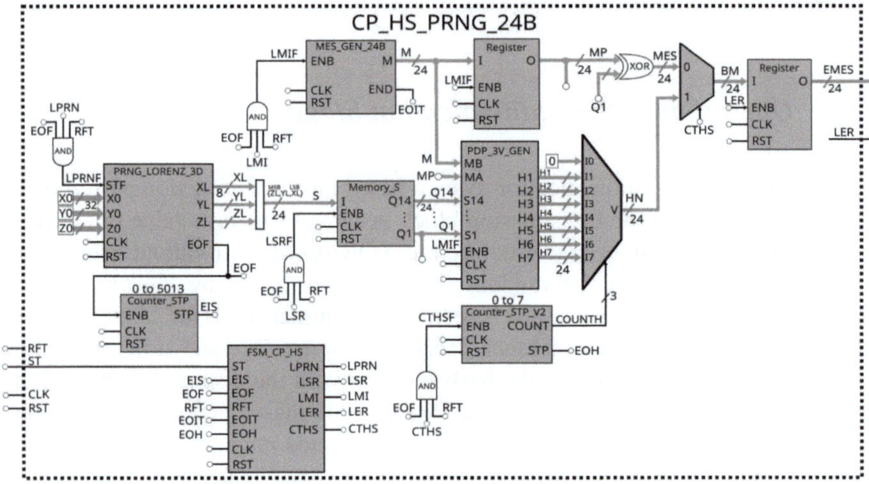

Fig. 7.3 Block description of the encryption and authentication system

send messages of the encrypted image and the hash value between the FPGA and a personal computer. This process of transmission is performed under the asynchronous interface RS232.

The block description of the transmitter of the encrypted and authentication parts is shown in Fig. 7.3. In this diagram one can see the chaos-based PRGN along other blocks such as counters, registers, memories, and a finite-state machine.

The block labeled as MES_GEN_24 provides a bus of 24 bits to transfer the messages that are formed from an RGB image of 256×256 pixels. Figure 7.4 shows the internal hardware description. The block consists of three ROMs and a counter with a stop pin. Each ROM stores a byte of a color of the image. With the three bytes a word of 24 bits is formed, and then this word is encrypted with the 24 bits generated by the chaos based PRNG. The block Counter_STP_V2 is similar to the block Counter_STP, having a difference that it provides the counting value as output. For an image of 256×256 pixels, the block counts from 0 to 65,536.

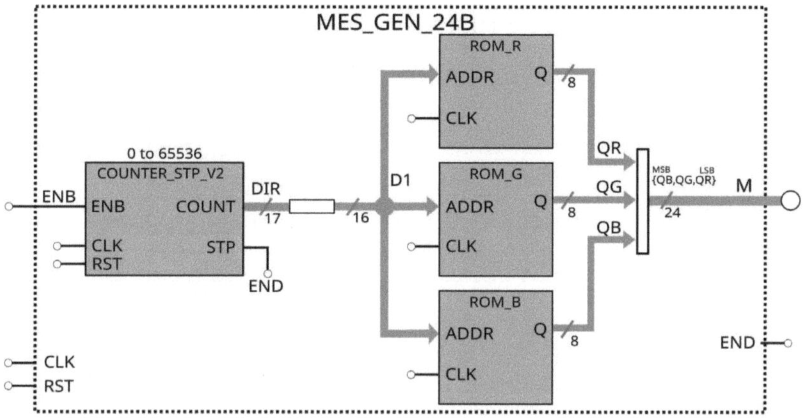

Fig. 7.4 Internal hardware description of the memory block for the RGB image

The diagram shown in Fig. 7.3 also embeds the block labeled as Memory_S, which is a memory composed of 14 PIPO registers in series connection. The function of this block is to store the time series *S* that come from the PRNG, which are required to calculate the Hash function by applying the pseudo dot product (PDP), as shown in Eq. (7.2). On the other side, the block PDP_3V_GEN is in charge of performing the mathematical operations of this equation. As inputs, the block receives two messages of 24-bit length and the 14 series *S*. The internal hardware description of this block is shown in Fig. 7.5.

The diagram shown in Fig. 7.5 is composed of seven blocks PDP_3V_H and a flip-flop type D, connected in type-T configuration. The blocks PDP_3V_H are in charge of performing the XOR operations, the multiplication, and the module operation, among the messages *M* and the respective series *S*, according to the variable *H* being calculated. The flip-flop type-T allows updating the values of each variable *H*, every two levels in high state of the input pin LMIF. The internal diagram of the block PDP_3V_H is shown in Fig. 7.6. Within the block PDP_3V_H, one can see the logic gates XOR, a multiplier, and an structure XOR-accumulator. The module operation is performed with the number 2^{24} to simply take the 24 LSBs from the generated 48 bits that are provided by the multiplier.

In the block diagram of higher hierarchy shown in Fig. 7.3, the encryption process occurs between the output Q1 of the block Memory_S and the pass message (MP) provided by the block MES_GEN_24.

On the other hand, the general functioning of the system is controlled by a finite-state machine (FSM) of Moore type, whose general description is explained in the following text:

1. EDO0: After the key is defined, the FSM activates the signal LPRNG (load PRNG) that executes the PRGN by a defined number of iterations to eliminate the transient behavior of the chaotic system. In this process the block Counter_STP

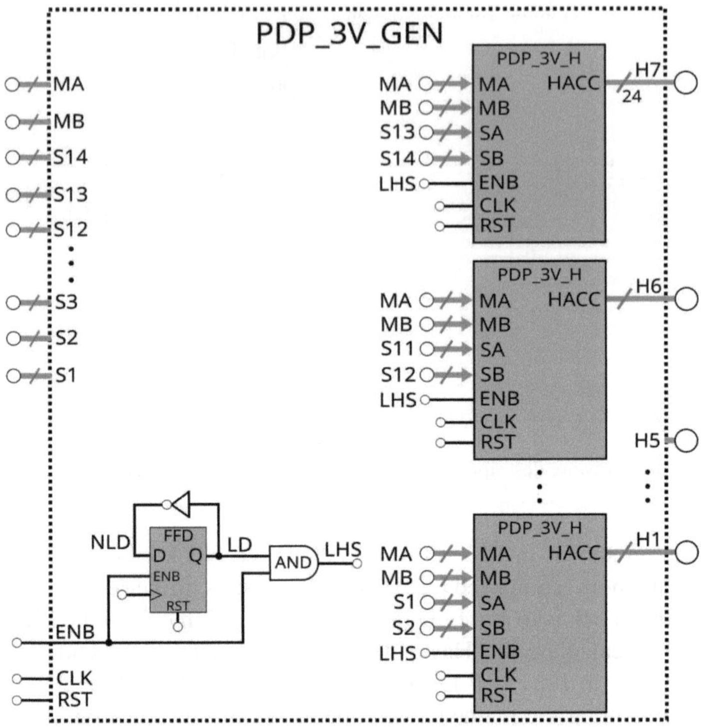

Fig. 7.5 Internal block diagram to get the hash value

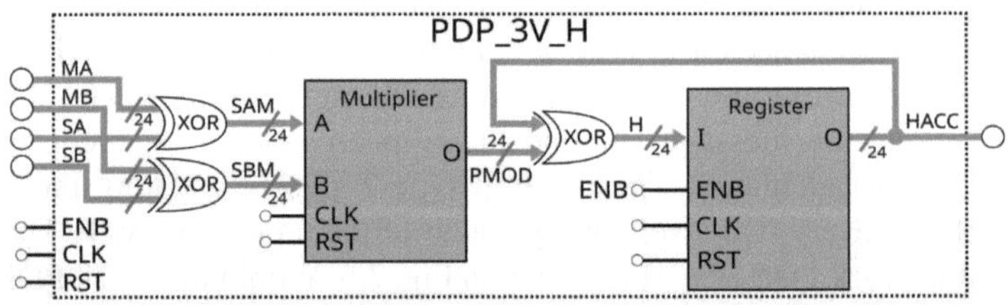

Fig. 7.6 Internal diagram description to calculate a variable *H*

also participates, which performs the counting. When the counting finishes, it activates the signal EIS (end initial series) in order that the FSM changes to EDO1. In addition, during this state, the block Memory_S stores the last 14 series of random numbers. Each value *S* is stored during the activation of the signals EOF (end of function) and LSR (load S registers).

2. EDO1: The FSM waits for the signal ST, to begin the encryption and authentication processes, and when ST is activated, the FSM goes to EDO2.
3. EDO2: The FSM activates the signals: LPRNG to generate a new times series *S*, LMI (Load_Memory_Image) to change the message of the block MES_GEN_24

and to execute the hash algorithm in the block PDP_3V_GEN, and LSR to update the memory of the series *S*. Each block is updated when the PRNG ends generating its value *S*. To go to EDO3, the signal EOF and RFT (ready for transmission) must be in a high logic state.

4. EDO3: The FSM activates the signal LPRNG to generate a new iteration and changes to EDO4 in one clock cycle (CC).
5. EDO4: The FSM activates the signal LER (load encryption register) in order that the PIPO register at the output stores the encrypted message. The address to the next state depends of the signal EOIT (end of image transmission). If this signal is in a high state, it means that the image has been totally encrypted and the FSM goes to EDO5; on the other hand, if it is in a low state, it goes to EDO2 and repeats the process until the image is encrypted.
6. EDO5: The FSM activates the signal CTHS (counter hash) to increment a counter (only one unity) and to change the input of the multiplexer that connects a variable *H* with the output. To change to EDO6, the input RFT and EOF must be in a high logic state.
7. EDO6: The FSM generates a stop of one CC before it goes to EDO7. During this state the signals LPRN and CTHS are activated.
8. EDO7: The FSM activates the signal LER to store a new value *H* in the output register and asks for the signal EOH (end of hash); if it is in a high level, the process ends in ED08, and if it is in a low state, it returns to EDO5 to repeat the process and to send all variables *H* through RS232. During this state the signals LPRNG and CTHS remain activated.
9. EDO8: In this state, the FSM waits for the signal reset to restart the transmission.

Figure 7.7 shows the block diagram for the transmission architecture by using the block CP_HS_PRNG_24B and one block for the interface RS232.

This system has the purpose of showing that a chaos-based PRNG is useful to develop encryption and authentication stages of RGB images. The system only represents an architecture for testing the algorithms and it is not a full protocol to provide security to the data during their transmission. In addition, this system takes advantage of the parallelism in the FPGA, because the hash function algorithm occurs simultaneously to the image encryption, where both processes employ the same series of random numbers *S*.

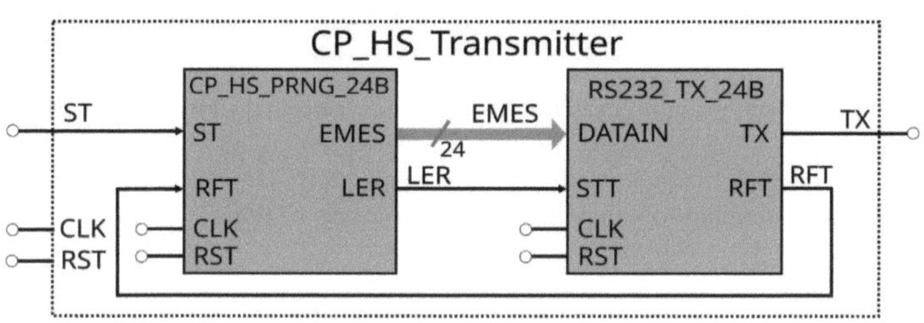

Fig. 7.7 Block diagram for the transmission system

A secure communication system can be improved by performing the transmission and reception in a wireless environment, where the decrypted image can be stored in a RAM module. Afterwards, the data can be sent to a device to be authenticated.

7.5 Verilog Simulation and FPGA Implementation Results

This section shows the experimental results of the encryption and authentication system synthesized into an FPGA. This is described in Fig. 7.7, by employing three initial conditions equal to 1, for the chaos-based PRNG, and performing 5,000 iterations to eliminate the transient behavior of the chaotic system. Figure 7.8 shows the RGB image of Lena having 256×256 pixels; from the left to the right the original, encrypted, and decrypted image can be appreciated. The tags for the authentication process are given in Table 7.1, in hexadecimal, which are provided by the transmitter and receiver blocks. As the image was not corrupted during its transmission, both tags are the same.

Figure 7.9 shows the simulation by using the software Active-HDL, where 7 values can be observed for the variables H, which correspond to the authentication tag given in Table 7.1.

On the other hand, when an image suffers an alteration or corruption in any pixel, both authentication tags will not be the same. Figures 7.10, 7.11 and 7.12 show three different images which has one altered pixel, replaced by white or black in a given position. Tables 7.2, 7.3 and 7.4 give the different hash values obtained from this alteration.

Table 7.5 shows the hardware resource consumption of the block transmitter synthesized on an FPGA Cyclone IV.

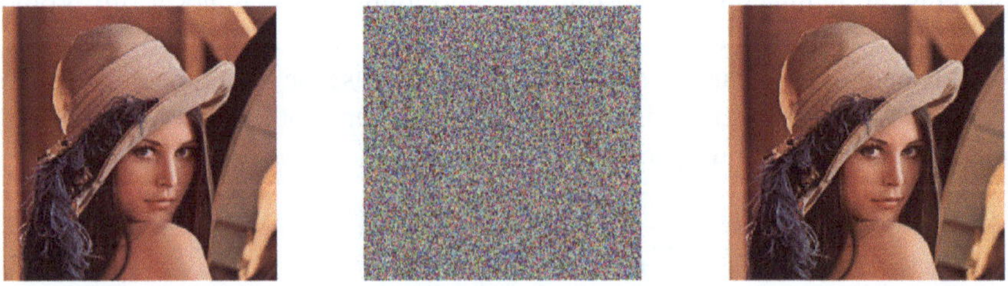

Fig. 7.8 Encryption/decryption process for Lena's image. Original (left), encrypted (center), and decrypted (right)

Table 7.1 Authentication tags obtained with the RGB image of Fig. 7.8

Device	168-bit hash value
Transmitter	0x480B76FEEB448E9F864F265A23BD064AD80A8395DE
Receiver	0x480B76FEEB448E9F864F265A23BD064AD80A8395DE

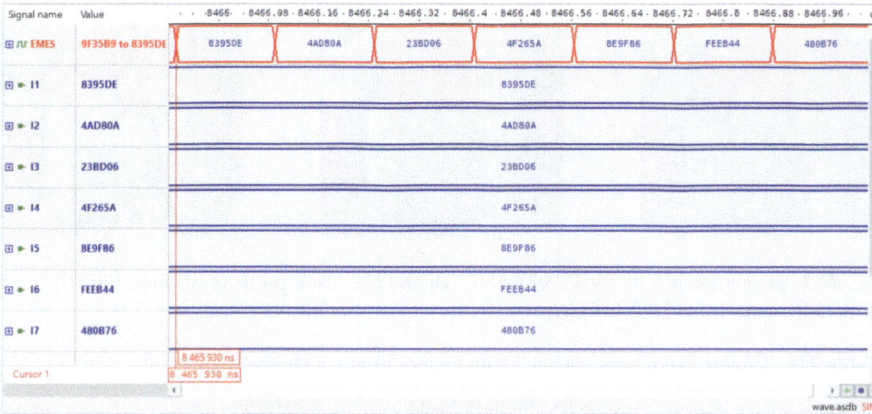

Fig. 7.9 Authentication tag provided by the software Active-HDL

Fig. 7.10 Lena images of 256×256 pixels, which have white pixels at position [64,64] (left), [128,128] (center), and [192,192] (right)

Fig. 7.11 Baboon images of 256×256 pixels, which have black pixels at position [96,96] (left), [128,128] (center), and [160,160] (right)

Fig. 7.12 Pepper images of 256×256 pixels, which have black pixels at position [80,80] (left), [128,150] (center), and [176,176] (right)

Table 7.2 Authentication tags obtained with images of Fig. 7.10

Image	168-bit hash value
Original	0x480B76FEEB448E9F864F265A23BD064AD80A8395DE
[64,64]	0xE7C93B4DBE7BD6CBF48D681F4578D9FCA535E4A214
[128,128]	0xA9A1E66EF62CBDF23EA8DBF2DEC65EF47742856B1E
[192,192]	0x01D5940304E45409846E6B806D2CC2C15B3074B084

Table 7.3 Authentication tags obtained with images of Fig. 7.11

Image	Hash value
Original	0xCB6C48A74490A05BC1EFA3AE83E78B441609A58CBB
[96,96]	0x3CE47834A3A01E9491313CE2AB12ABAA2E5D30BBF7
[128,128]	0xE8EDA229E00814E5E99B027691EB070A7AA9956165
[160,160]	0x4E65489E19900531C10E0DAE2E620B93C889AC393B

Table 7.4 Authentication tags obtained with images of Fig. 7.12

Image	Hash value
Original	0x25BA6EA710C5490E01B8566553CB523E9072A8BD18
[80,80]	0x6D81565333AB03396F382D8B82E44EF8023C5054A4
[128,150]	0x2139A6878577ECE58C221A620B122F2E54469F1BC2
[176,176]	0x85AFA49A9955B950751518FFA12AD64FAA062D16B2

7.6 System for Text Authentication and Encryption

The secure communication system for the encryption and authentication shown in Fig. 7.3 can also be used to process messages from greyscale images and text. This section uses as case study, the text given in the abstract of Chap. 1. In this case, the binary values of each symbol in the text must be stored. That values correspond to ASCII characters and are stored in the memory block MES_Gen_24B shown in Fig. 7.4. For this case study, the encryption is performed through the operation XOR among three contiguous characters and one sequence of time series S of 24 bits generated by the chaos-based PRNG. In the same manner as for the images, the hash

Table 7.5 FPGA Cyclone IV EP4CGX150DF31C7 hardware resources of Fig. 7.3

Resources	Used	Available
Logic elements (LE)	2,221	149,760
Combinational functions	2,033	149,760
Dedicated logic registers	1,332	149,760
Memory bits	1,572,864	6,635,520
Embedded multiplier 9-bit	58	720
Maximum frequency (MHz)	66.89	–
Latency at 50 MHz (ns)	120	–
Throughput at 50 MHz (Mbps)	200	–
Efficiency (Kbps/LE)	0.090	–

function is executed performing XOR operations and multiplications in an iterative way. The implementation performed in this section, the number of stored characters in the memory, is established before the execution of the system. Recall that an FPGA implementation of an authentication of encrypted data as images and text requires the design of control blocks to determine the quantity of data in the message and then to establish the number of iterations of the hash function algorithm.

Figure 7.13 shows the encryption and decryption of the text in the abstract given in Chap. 1 in this book. The text has 792 characters and one can appreciate that letter A in the word Abstract, was modified during its transmission, in such a way that in the recovered message, it appears as letter O. In this manner, as the decrypted text is not the same as the original one, both authentication tags are different, implying that the text has been altered during its transmission. Table 7.6 shows the authentication tags got with the original and altered text in letter A.

The authentication tests were also performed through a communication between an FPGA and a personal computer using the interface RS232, where the text is stored in the memory block MES_Gen_24B, and the processing was performed using the architecture shown in Fig. 7.3.

The input text for the hash function and the encryption process can be coded in the positional enumerated system BASE64 to use a text format that can be employed in text-based protocols.

7.7 Conclusions

This chapter described the FPGA synthesis of an authentication system that works with encrypted images by using chaotic time series or chaos based PRNGs. It was mentioned that a hash algorithm can increase its resistance to collision by employing the modulo operation with long prime numbers. However, the division algorithm is complex to describe under Verilog and require a considerable processing time, so that this book used a power of 2.

Abstract The design and implementation of random number generators(RNGs) have been widely studied during the last years to improve applications that require random sequences. The RNGs can be classified as true (TRNGs) or pseudo (PRNGs), and either or both can be designed by using different algorithms or hardware. This section shows the basics of PRNGs and introduces the ways they are implementation by using chaotic maps, chaotic systems and fractional-order chaotic oscillators. The goal of the book is devoted to provide guidelines for the block descriptions of chaotic maps and systems, their Verilog descriptions, and their FPGA implementation. This chapter provides a general overview of the application of RNGs, their design using chaotic systems, and their hardware implementations.

(a)

Ð§c°T□¸©=-þG®Ü£BQêÙ@!ZìÓ□¨□□à□(Oîb+á□□9[¢ÈjÈ&Kæéûj□olÔ²å:rû,ß¤Íðý□[□□~àë
Äk□□_,ã½Ñ□n3{\□L²ð□T[¨□JýÙs□□:□!□□]X¡×~þP¡p□DÎ0ï°ÇB□kÝ÷¼ay÷gÂÐ▲V□
□`ë□ÍÕ0□·þ□î□µâ²□!Û▲□Zá¡ñûCæ=□°ÅAÆ□□ z2þ¨□þ×Õ□
È¢Øo[¡õ4þ1ÛV□□/□1^`d,B□«ôí£¤`□Ö□Ä3ã«T□vq□□Ú&□(□ï□□Þ»qDÒ□?□ê1ØÔ$O□-Î□Ç
÷□²nh□f□:É/¸ÒròÓ6¼¸□håþe□[4¥ ¤8ka□{Ù□3ù0åg□Æ}Ä□üté▲
sJÁ□s□K¿M%PÅkW□Îõ¸Ê□÷ÞÃ\□□0□ïúO+«□´□
9÷O½úÕ□ Xð□h□□8-¸j.ÈE□<K»□¸□484')[&□½®'`e□îæøs´(¶□□□□□¤áù¬□□IE□×□lä>.?
Þ}$□Oã□spf□9S¨çÆÍ
6□Û□□ûAÅuò©'äÈ

(b)

Obstract The design and implementation of random number generators(RNGs) have been widely studied during the last years to improve applications that require random sequences. The RNGs can be classified as true (TRNGs) or pseudo (PRNGs), and either or both can be designed by using different algorithms or hardware. This section shows the basics of PRNGs and introduces the ways they are implementation by using chaotic maps, chaotic systems and fractional-order chaotic oscillators. The goal of the book is devoted to provide guidelines for the block descriptions of chaotic maps and systems, their Verilog descriptions, and their FPGA implementation. This chapter provides a general overview of the application of RNGs, their design using chaotic systems, and their hardware implementations.

(c)

Fig. 7.13 Encryption/decryption of the text in the abstract in Chap. 1 by using the architecture shown in Fig. 7.3: (**a**) original text, (**b**) encrypted text, and (**c**) decrypted text, where the first letter was changed from A to O

Table 7.6 Authentication tags obtained with the text given in Fig. 7.13

Text	Hash value
Original	0x621F7981AAEF34F257ACA2A2C628BFC485F5AB1DCD
Decrypted	0xA464F9A593956CC46F4AB522FA27716D7EA38FFFA1

The Verilog description of the hash algorithm can be used with other chaos-based PRNGs with a higher or lower number of bits. Simply, one must adjust the size of the buses of the message and the time series of the PRNG. One must also modify the way the message is generated from the bytes representing the colors of the image.

This authentication of encrypted images system can also be applied to process text and black and white images.

References

1. E. Tlelo-Cuautle, A.D. Pano-Azucena, O. Guillén-Fernández, A. Silva-Juárez, *Analog/Digital Implementation of Fractional Order Chaotic Circuits and Applications* (Springer, Berlin, 2020)
2. E. Tlelo-Cuautle, L.G. De La Fraga, O. Guillén-Fernández, A. Silva-Juárez, *Optimization of Integer/Fractional Order Chaotic Systems by Metaheuristics and Their Electronic Realization* (CRC Press, Boca Raton, 2021)
3. J. Hwang, G. Kale, P.P. Patel, R. Vishwakarma, M. Aliasgari, A. Hedayatipour, A. Rezaei, H. Sayadi, Machine learning in chaos-based encryption: theory, implementations, and applications. IEEE Access **11**, 125749–125767 (2023)
4. L.M. Pecora, T.L. Carroll, Synchronization in chaotic systems. Phys. Rev. Lett. **64**(8), 821 (1990)
5. L.G. De la Fraga, C. Mancillas-López, E. Tlelo-Cuautle, Designing an authenticated hash function with a 2d chaotic map. Nonlinear Dyn. **104**(4), 4569–4580 (2021)
6. O. Martínez-Fuentes, J.D. Díaz-Muñoz, A.J. Muñoz-Vázquez, E. Tlelo-Cuautle, G. Fernández-Anaya, I. Cruz-Vega, Family of controllers for predefined-time synchronization of lorenz-type systems and the raspberry pi-based implementation. Chaos Solit. Fract. **179**, 114462 (2024)

Index

A
Active-HDL, 104, 121, 122, 130, 131, 137, 140, 141, 164, 169, 170, 186, 187
ASCII characters, 88, 188
Assign, 37, 38, 167
Authentication, 3, 12, 18, 32, 35, 37, 172, 177–190

B
Binary sequences, 2, 4, 6, 8, 12, 28, 83–85, 88, 95, 97, 98, 102, 108, 110, 114, 116, 118, 132, 167, 177

C
Chaotic attractor, 11, 28, 73, 100, 101, 111, 123, 125, 126, 128, 134, 137, 141, 142, 146, 166, 170–172
Chaotic map, 1–12, 32, 35–80, 83, 85, 95–143, 177–179
Chaotic systems, 1–12, 17–32, 35–80, 85, 88, 95–174, 177–179, 183, 186
Chaotic time series, 10, 18, 20, 24, 32, 100–102, 105, 106, 121–124, 126, 127, 130–132, 135, 136, 139–141, 167, 170, 172, 177, 189
Clock cycles (CCs), 39, 41, 46, 50, 52, 54, 61, 63, 71, 75, 78, 102, 107, 123, 124, 130, 132, 133, 136, 140, 157, 161, 162, 167, 168, 185
Computer arithmetic, 35, 36, 40, 65, 66, 71, 72, 74, 96, 101, 102, 109, 135, 148, 164, 169
Correlation, 98, 99, 110, 113–116, 118, 119
Cryptography, 5–7, 11, 12, 35

D
Dieharder test suite, 91–92, 177
Digital-to-analog (DAC) converter, 105, 170

E
Eigenvalue, 25–27
Encryption, 3, 5–7, 12, 18, 27, 28, 37, 145, 172, 177–190
Equilibrium points, 11, 18, 25–27

F
Field-programmable gate array, 3, 20, 35
Finite state machine (FSM), 39, 50, 54–56, 62, 74–76, 156–158, 161, 162, 182–185
5D chaotic system, 25, 27, 95–143, 177
Fixed-point format, 12, 35, 36, 43–46, 71, 72, 74, 78, 97, 98, 101, 109, 118, 126–127, 135, 136, 138, 140, 148, 151, 157, 164, 169, 171, 179
Floating point number, 88–90, 92, 98, 107, 114–117
Forward Euler (FE) algorithm, 117

L. G. de la Fraga et al., *Random Number Generators*,
https://doi.org/10.1007/978-3-031-82865-2

4D chaotic system, 27, 95–143
Fractional-order differential equations (FODEs), 35, 37, 45, 47, 65, 145, 146
Fractional-order memristive system, 29, 148, 150–171
Fractional-order systems, 21, 146

G
Grünwald-Letnikov method, 20, 146

H
Hardware description language (HDL), 24, 36, 37, 101
Hash function, 5, 6, 178–183, 185, 189
Hash tag, 180, 181
Hexadecimal, 15, 104, 170, 186
Hidden attractor, 18, 23–28

I
Image processing, 185, 186, 188, 190
Initial condition, 2, 3, 17, 18, 24, 27, 28, 32, 35, 42, 65, 73, 74, 98–102, 109, 110, 113–115, 117–120, 126, 129, 134, 148, 150, 167, 168, 178, 179, 186
Instantiation, 37, 38, 60, 64, 69, 72, 80, 121, 159, 177

L
Least significant bits (LSBs), 95, 96, 98, 101, 102, 106, 108, 110, 112, 114, 116, 118, 123, 126, 132, 135, 139, 140, 142, 143, 167, 171, 172, 183
Lyapunov exponent, 17, 18, 24–26, 97, 177

M
Memory cell, 56, 58, 60, 78
Memristor, 10–12, 25, 29–31, 35, 145–147, 177
Most significative bit (MSB), 127
Multi-step method, 19

N
National Institute of Standards and Technology (NIST), 2, 4, 6–8, 10, 12, 28, 32, 36, 83–88, 91, 92, 98, 99, 101, 102, 110, 113–118, 126, 129, 141, 145, 167, 177
Numerical method, 1, 12, 17–32, 35, 37, 45, 47, 65, 69, 71, 145, 147, 177

O
1D map, 4, 95, 97
One-step method, 19, 24, 35
Ordinary differential equations (ODEs), 17, 18, 23, 24, 32, 35, 37, 45, 47, 65, 117, 125, 135, 145, 146
Oscilloscope, 105, 106, 123, 132, 139, 141, 170–172

P
Parallel input parallel ouput (PIPO), 38–44, 46, 49, 51, 58, 60, 62–64, 67, 80, 119, 130, 150, 151, 157, 159, 167, 183, 185
Pseudo dot product, 180–181, 183
Pseudo random number generators (PRNGs), 1–12, 20, 27, 28, 31, 32, 35–37, 73, 80, 83–92, 95–174, 177–185, 189

Q
Quartus-II compiler, 36

R
Random memory access (RAM), 56–58, 89, 186
Randomness, 2–4, 12, 20, 32, 83–87, 89, 92, 95, 96, 98, 101, 110, 114, 116, 118, 126, 141, 145, 167, 177
Random number generators (RNGs), 2, 4, 5, 12, 83, 90–92
Read only memory (ROM), 57, 58, 78, 157, 158, 161, 182
Register, 38–44, 46, 49–51, 58, 60–64, 67, 74, 80, 107, 119, 124, 130, 133, 141, 145, 150, 151, 156, 157, 159–162, 167, 168, 171, 182–185, 189
Register based memory (RBM), 58–60, 159–161

S
Self-excited attactor, 18, 23–28
Serial peripheral interface (SPI), 105, 171

Short memory length, 147
Signed, 37, 43–46
Single constant multipliers (SCMs), 45–47, 65–69, 71, 72, 74, 127, 136, 151, 154, 155
Step-size, 20, 23, 28, 32, 65, 67, 117, 126, 134, 148
Synchronization, 3, 178, 179

T
TestU01 suite, 87–90, 92
3D chaotic system, 24, 25, 95–143
Throughput, 11, 106, 107, 123, 124, 132, 133, 140, 141, 143, 171, 189
Time delay, 29–32, 35, 146–172, 177
2D map, 11, 64–80, 85, 88, 97–112, 120, 123, 132, 136, 139, 140
2's complement, 37, 40, 42

V
Verilog description, 28, 35–80, 98, 102–105, 119–123, 127–132, 135–139, 150–164, 167–170, 181–186, 190

W
Wire, 37, 46, 154

X
XOR operation, 95, 178–181, 183, 189

MIX
Papier aus verantwortungsvollen Quellen
Paper from responsible sources
FSC® C105338

If you have any concerns about our products,
you can contact us on
ProductSafety@springernature.com

In case Publisher is established outside the EU,
the EU authorized representative is:
Springer Nature Customer Service Center GmbH
Europaplatz 3, 69115 Heidelberg, Germany

Printed by Libri Plureos GmbH
in Hamburg, Germany